AF889609

The S-100 and Other Micro Buses

Second Edition

by
Elmer C. Poe
and
James C. Goodwin II

Howard W. Sams & Co., Inc.
4300 WEST 62ND ST. INDIANAPOLIS, INDIANA 46268 USA

Copyright © 1979 and 1981 by Elmer C. Poe
and James C. Goodwin II

SECOND EDITION
SECOND PRINTING—1982

All rights reserved. No part of this book shall be reproduced, stored in a retrieval system, or transmitted by any means, electronic, mechanical, photocopying, recording, or otherwise, without written permission from the publisher. No patent liability is assumed with respect to the use of the information contained herein. While every precaution has been taken in the preparation of this book, the publisher assumes no responsibility for errors or omissions. Neither is any liability assumed for damages resulting from the use of the information contained herein.

International Standard Book Number: 0-672-21810-0
Library of Congress Catalog Card Number: 81-50561

Printed in the United States of America.

Preface

Everyone's computer seems to shrink with age. This shrinkage is usually first apparent in the memory as what was, initially, an enormous memory becomes very confining. The addition of another memory board or two temporarily solves the problem. Also, at first, the lack of hardcopy seems almost a blessing but, with every passing clock cycle, the need for a written record becomes more important. A printer interface board is added in hopes that it will do the trick. However, there are a host of expansion boards available for expansion and the present system always seems to be wanting and waiting for more and more. The key to system expansion is the *bus* through which the processor communicates with the present and future system components.

This book is about buses. A general discussion in Chapter 1 is used to acquaint the reader with bus basics. Then, the mechanical data, pinout designations, and bus signal definitions of the 21 most widely used bus systems are listed and explained in the following chapters so that the reader may better understand the bus of his or her system, or evaluate the bus of another system, or plan the interface of one bus to another. Expansion and interface boards are described and explained so that a greater use of the bus system can be obtained. Many photographs and pinout drawings are shown so that a better understanding of microcomputers, interface boards, and *buses* can be realized.

Since the first edition of this book was published, personal microcomputers have changed. Most are now offered as a complete package that contains everything necessary for immediate use in BASIC. Although many do not contain the traditional motherboards, the bus is still important for expansion. In these systems the expansion bus is generally brought out to an edge connector.

The industrial microcomputer still contains a motherboard which must be populated with selected boards to do the particular job at hand. Such boards, which include CPU, memory, and i/o boards, have to be interconnected through a system bus.

We would like to express our appreciation to the many manufacturers who participated in the preparation of this book by sharing their schematics and system descriptions. Thus they have helped the reader to better understand and use their products. We also want to give special thanks to Debbie for her meritorious service in preparing the manuscript.

ELMER C. POE
JAMES C. GOODWIN II

To our wives, Debbie and Sue

Contents

CHAPTER 7

CHAPTER 8

CHAPTER 9

CHAPTER 10

CHAPTER 11

CHAPTER 12

CHAPTER 13

CHAPTER 14

CHAPTER 15

CHAPTER 16

CHAPTER 17

CHAPTER 18

CHAPTER 19

CHAPTER 20

CHAPTER 21

CHAPTER 22

CHAPTER 23

CHAPTER 24

CHAPTER 25

APPENDIX

CHAPTER 1

Bus Structures

After the introduction of the Altair 8800 in early 1975, interest in microcomputers by hobbyists began to grow rapidly. In those early days, getting a system set up and running was often a formidable challenge. Yet, even then, the need for system expansion was quickly felt. Companies producing Altair-compatible memory boards soon appeared and flourished. By the end of the year the Altair had become a leader with over 50 companies producing plug-compatible boards for the "Altair" or S-100 bus. During this period several other successful systems, including the SWTP 6800, KIM-1, and IMSAI, were introduced. Some of these systems used the Altair bus but many did not. Since then, many systems and buses have appeared. Memory, serial and parallel i/o, video and graphics display, analog i/o, voice systems and recognition, music synthesis, and a host of other boards are available for these systems. Figs. 1-1 through 1-9 show some typical systems and plug-in boards.

The Commodore PET shown in Fig. 1-1 is a complete microcomputer system in one unit. It has its own bus. The Atari 800 illustrated in Fig. 1-2 is also a complete microcomputer system with its own bus. The AIM 65 computer shown in Fig. 1-3 is a complete system that can be expanded via the bus connector at the rear of the board. The Heathkit H8 uses Heath's Benton Harbor Bus. It is complete as shown in Fig. 1-4 using keypad and octal displays or it is usable with other i/o devices.

Fig. 1-5 shows the Vector 1, an S-100-based system that relies entirely on external i/o devices. An S-100 serial communications board is illustrated in Fig. 1-6, while a Heath H8-2 parallel interface board is shown in Fig. 1-7. Notice the bus connector on the right side of the H8-2 board. Fig. 1-8 shows the Percom floppy disk and controller for SS-50 systems, while Fig. 1-9 illustrates the video digitizer board. It is using the SS-30 connector in an SS-50 system.

Courtesy Commodore Business Machines, Inc.

Fig. 1-1. PET, a complete microcomputer system.

The bus of a system is the selection and arrangement of its signal and power lines on a connector for distribution to other boards. Bus structure becomes important when system expansion is considered, since the needed signals must be available. Careful examination of the bus of a system should be made before a system is purchased. The most popular buses will be discussed in detail.

ADDRESS LINES

Regardless of the microprocessor used in a system or the number of lines on its bus, the address, data, and control signals must be available. The address lines are used by the processor to indicate to memory and other peripherals the location with which it wants

Courtesy Atari, Inc.

Fig. 1-2. Atari 800 computer.

Courtesy Rockwell International

Fig. 1-3. S-100 extender board.

Courtesy Heath Co.

Fig. 1-4. Heathkit H8 digital computer mainframe.

Courtesy Vector Graphic, Inc.

Fig. 1-5. Vector 1 is an S-100–based system.

Courtesy DCHayes Associates

Fig. 1-6. An S-100 serial communications board.

to communicate. The 8080, 8088, Z-80, 6800, 6809, and 6502 have 16 address lines, generally labeled A0–A15, that are divided into two 8-bit bytes. Lines A0–A7 comprise the low-address byte and lines A8–A15 comprise the high-address byte. Not all address lines may be used by any one board. The 8086, 68000, Z-8000, LSI-11, and TI9900 are 16-bit processors that may contain up to 24 address lines.

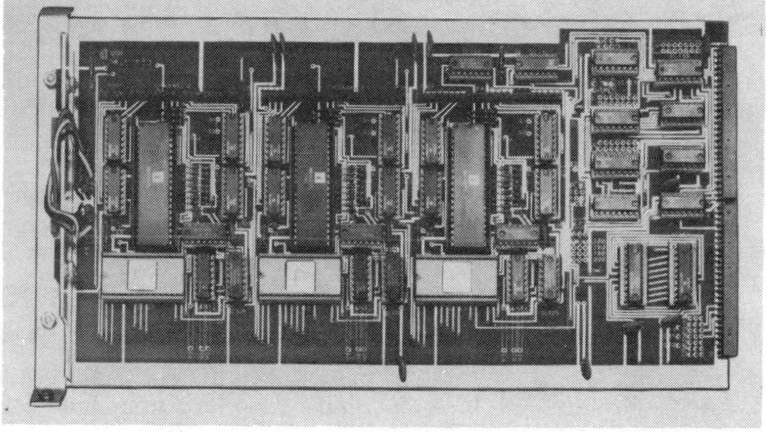

Courtesy Heath Co.

Fig. 1-7. Heath H8-2 parallel interface board.

Courtesy Percom Data, Inc.

Fig. 1-8. Percom floppy disk and controller for SS-50 systems.

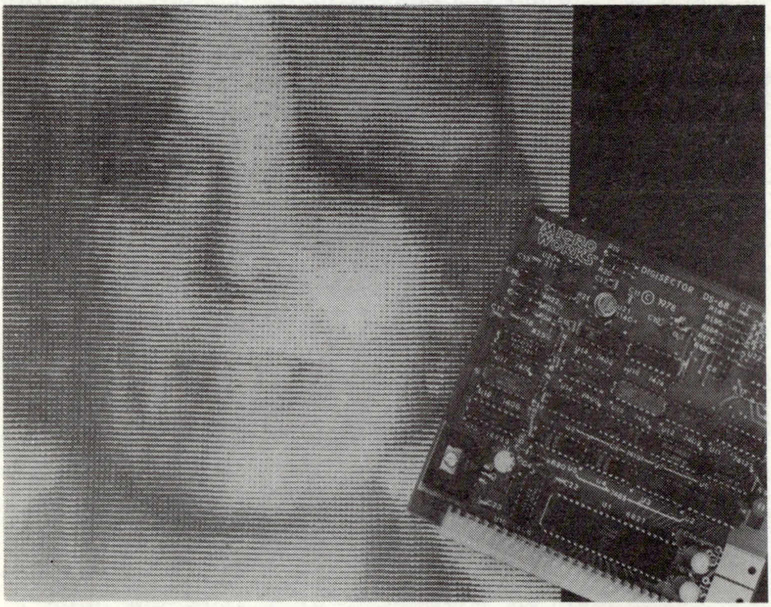

Courtesy The Micro Works

Fig. 1-9. Video digitizer board using the SS-30 connector in an SS-50 system.

DATA LINES

The data lines, usually designated D0–D7 on 8-bit buses and D0–D15 on 16-bit buses, carry instructions and data between the processor and all the peripherals, including the memory. All processors have bidirectional data lines which carry information both into and out of the processor. The direction of information flow on these lines is usually under processor control. All buses, except the S-100 and the Digital Group buses, maintain the bidirectional data lines. The S-100 splits the data lines into eight data-input and eight data-output lines. The Digital Group bus splits them into eight memory-data in, eight memory-data out, eight i/o in, and eight i/o out.

CONTROL LINES

The control lines coordinate the operation of all system components. Most buses include a master clock line, often $\phi 2$ on the processor, that indicates to peripherals that the address placed on the bus is settled and valid. The master clock line also indicates valid data is on the data lines during a write operation and that the processor can accept data placed on the data lines during a read operation. The 8080 and Z-80 buses may include an M1 line that signals the beginning of an op-code fetch and a sync line that signals the beginning of each machine cycle.

Direction of data flow on the data lines is determined by one or more processor outputs. The 6800 and 6502 systems use a single read/write line to control data direction. When the line is high a read operation is indicated, when low a write operation will occur. This 8080- and Z-80-based systems use separate read and write lines and they separate memory from other i/o with two sets of control lines. The memory-read and memory-write lines control direction of data flow during memory operations. The i/o read and write lines control data direction during i/o operations. Most buses include a ready line that can be used by slow memory or i/o to momentarily halt the processor until valid data can be accessed.

Most buses include reset lines. Although a few of these are connected directly to the processor's reset input, many are not. Some are outputs of a reset pulse generator that can be used to reset peripherals. Others are inputs to circuitry that reset the operating system, often through a nonmaskable interrupt input, without resetting the processor.

Interrupt inputs are common to all systems. These inputs allow infrequent or important peripherals to get the attention of the system. A maskable interrupt input line can be ignored by the processor on direction from the program. A nonmaskable input cannot be ignored;

it must be serviced by the processor. Vectored interrupt inputs feed an interrupt controller that assigns priority if more than one interrupt request should occur at one time. The number and type of interrupt inputs vary widely among buses. The 8080 and Z-80 buses may include status lines that indicate the condition of the maskable input or acknowledge the processor's recognition of an interrupt request.

Some micro buses are designed to allow more than one device to gain control of the bus lines. A system may have more than one processor or a variety of intelligent devices, i.e., disk controllers, arithmetic and i/o processors, etc., on the bus. Generally, any device that is capable of controlling bus operation is called a *bus master*. Devices that operate on the bus but cannot control it, i.e., memory, i/o ports, etc., are generally called *slaves*. A system will always have at least one master and one slave, and it may have several of each.

All processors require at least +5 volts dc. The 8080 also needs −5 and +12 volts dc. PROMs and communications devices may need −5, −9, or −12 volts. Most buses supply these voltages as filtered but unregulated +8 and ±16 volts, with regulation occurring on the individual boards.

SYSTEM EXPANSION

Expansion of some systems occurs via a motherboard, as shown in Fig. 1-10. A motherboard is a printed-circuit board that distributes bus signals to several connectors which mate with system boards. Motherboards are an integral part of some systems but single-board computers may require their addition to allow for expansion. Adding a board to a system is not as simple as just purchasing it and plugging it in. A space must be available on the motherboard; indeed, a motherboard must be present. The power supply of the system must produce the necessary voltages and be able to provide the extra current required by the new board. The power requirement may become critical in a system that is being expanded.

An extender board, shown in Fig. 1-11, is used to raise one system board above the remaining boards to facilitate testing and troubleshooting.

The microprocessor board of the system must be able electrically to drive the devices on the new board. The processor has very limited drive capabilities, so most systems pass the signals of the processor through buffer chips. These devices provide little load on the processor and are capable of driving up to 100 other devices. Without buffering, the processor can be fatally damaged. However, if a system does not buffer the signals of the processor, a buffer board may have to be added. Also, the speed of the new board must be compatible with the system. Memory and peripherals may not be

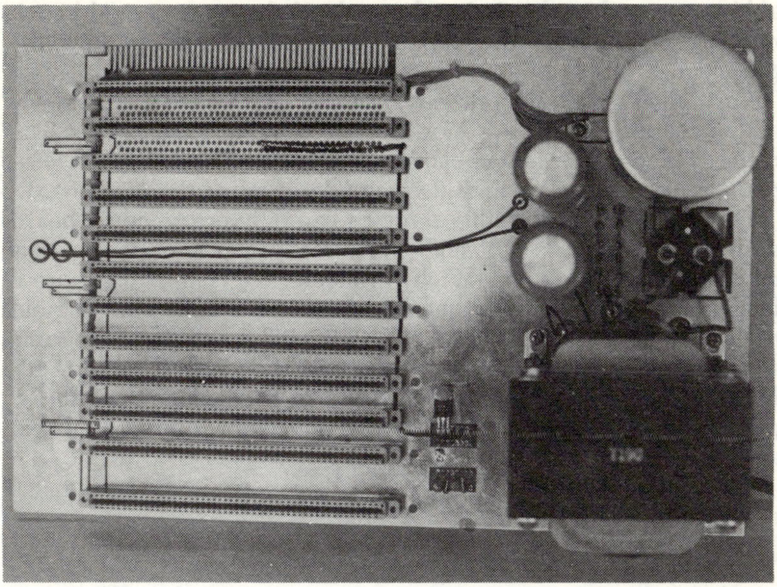

Courtesy Quay Corp.

Fig. 1-10. A 12-slot expandable S-100 motherboard.

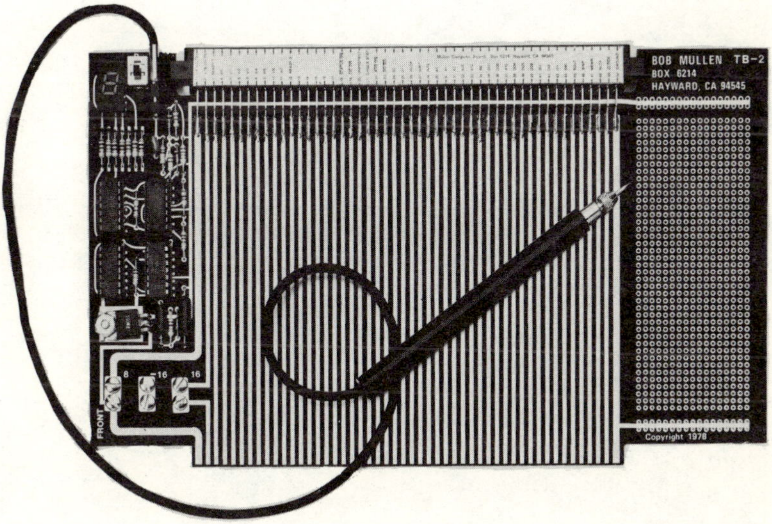

Courtesy Mullen Computer Products

Fig. 1-11. S-100 extender board.

able to respond to the timing demands of the processor. Although most systems include the "READY" input to temporarily slow the processor, this may be unacceptable.

Care should be taken when assigning addresses to the newly added boards. New boards must be placed in unused locations. The memory map of the system can be examined to determine the locations of existing boards and possible addresses for the new boards.

Finally, plug compatibility must be carefully examined. A board designed for one bus will not plug directly into another, but bus interface boards are commercially available. Yet, compatibility problems exist between boards that are built for the same bus. The old rule of buying the processor and all additional boards from the same company no longer holds true. The pinout and signal definitions for both the system and the new board should be examined before the purchase is made. Specifications of the most popular buses are listed and explained in the following pages.

CHAPTER 2

The S-100 Bus

The S-100 bus was introduced by MITS, Inc., on their 8080-based Altair kit. In fact, many of the bus lines are signals which are generated by the 8080 processor. The S-100 has become the most popular bus, if not the industry standard, in use by hobbyists today. Fig. 2-1 shows an S-100 microcomputer system which includes a keyboard, video monitor, and dual floppy disk. There are more than 100 boards being advertised as directly compatible with the S-100 bus. These include a multitude of memory boards, serial and parallel interface boards, floppy disk controller boards, video boards, music synthesizer boards, and speech recognition boards. Some examples are shown in Figs. 2-2 through 2-5.

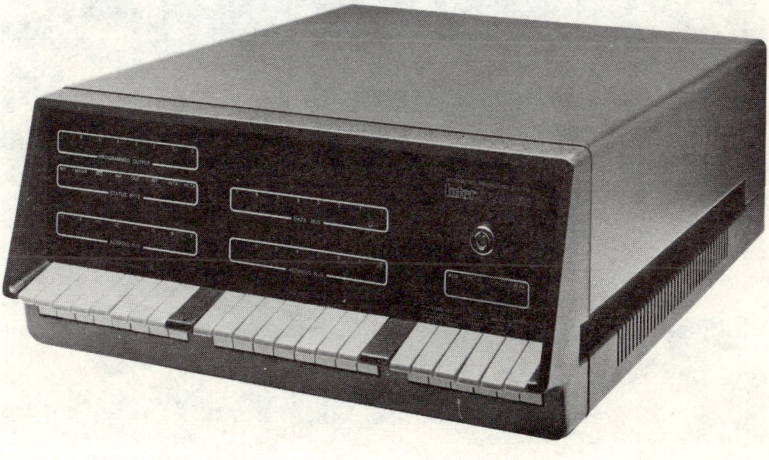

Courtesy Intersystems

Fig. 2-1. Intersystems S-100 mainframe.

Courtesy Quay Corp.

Fig. 2-2. A memory board (8K static RAM).

The S-100 circuit board (for example, Fig. 2-2) has overall dimensions of 5.3 by 10 inches (13.4 by 25.4 cm). It fits into a 100-pin edge connector (50 on each side, spaced 0.125 inch, or 0.317 cm, apart)

Courtesy Processor Technology Corp.

Fig. 2-3. A floppy disk with control boards.

Courtesy Xitex Corp.

Fig. 2-4. Stand-alone video board. Only +7 Vdc and GND on the bus are used.

on a motherboard (Fig. 2-6). The pins are numbered from left to right as shown in Fig. 2-2 with pins 1–50 on the component side and pins 51–100 on the opposite side. The bus supplies unregulated voltages to the boards; therefore, all boards must contain the necessary regulators. The pins are offset toward onc end so that the boards cannot be inserted backwards. The S-100 pinout designations are listed in Fig. 2-7.

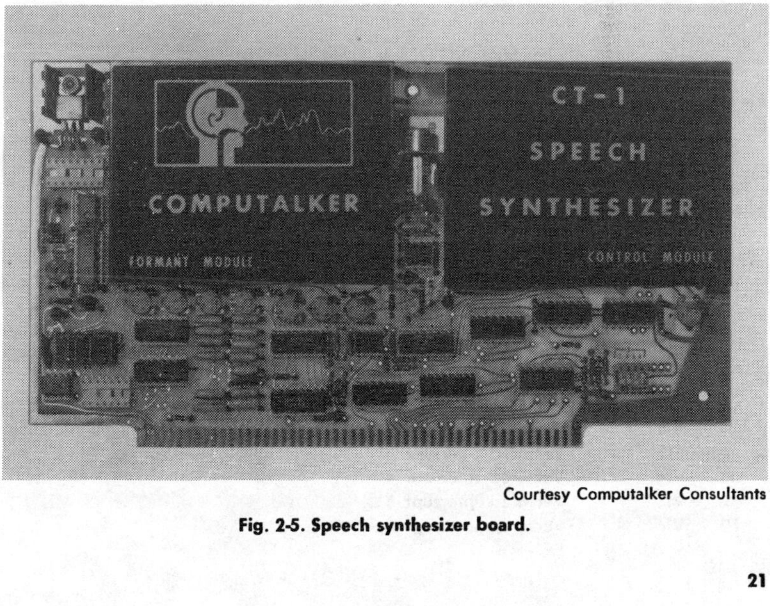

Courtesy Computalker Consultants

Fig. 2-5. Speech synthesizer board.

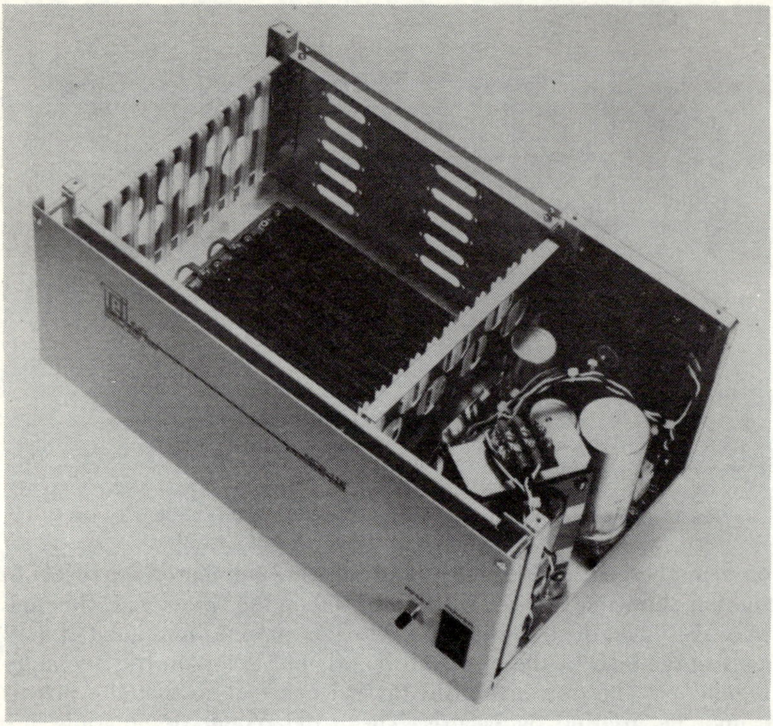

Courtesy CMC Marketing Corp.

Fig. 2-6. S-100 mainframe including chassis, 12-slot motherboard, and power supply.

S-100 SIGNAL DEFINITIONS

Pin No.	Signal	Definition
1	+8 V	Unregulated input to +5-V regulators.
2	+16 V	Positive unregulated voltage supply.
3	$\overline{\text{XRDY}}$ (READY 1)	External ready—ANDed with $\overline{\text{PRDY}}$ (pin 72) and connected to READY on the 8080. If $\overline{\text{XRDY}}$ and/or $\overline{\text{PRDY}}$ are pulled low, the CPU will enter a wait state until both are high. $\overline{\text{XRDY}}$ is often used as a front panel control and can allow single stepping. $\overline{\text{PRDY}}$ usually is used to signal when valid data from slow memory is on the bus.
4	VI0	Vectored interrupt 0—A vectored interrupt system is used when very fast multiple interrupt response is required and is implemented with a special circuit card.
5	VI1	Vectored interrupt 1.
6	VI2	Vectored interrupt 2.
7	VI3	Vectored interrupt 3.
8	VI4	Vectored interrupt 4.
9	VI5	Vectored interrupt 5.
10	VI6	Vectored interrupt 6.

Pin	Signal	Signal	Pin
51	+8 V	+8 V	1
52	−16 V	+16 V	2
53	$\overline{\text{SSW DSB}}$	$\overline{\text{XRDY}}$	3
54	$\overline{\text{EXT CLR}}$	VI0	4
55	*	VI1	5
56	*	VI2	6
57	*	VI3	7
58	*	VI4	8
59	*	VI5	9
60	*	VI6	10
61	*	VI7	11
62	*	*	12
63	*	*	13
64	*	*	14
65	*	*	15
66	*	*	16
67	*	*	17
68	MWRITE	$\overline{\text{STAT DSB}}$	18
69	$\overline{\text{PS}}$	$\overline{\text{C/C DSB}}$	19
70	PROT	UNPROT	20
71	RUN	$\overline{\text{SS}}$	21
72	$\overline{\text{PRDY}}$	$\overline{\text{ADD DSB}}$	22
73	$\overline{\text{PINT}}$	$\overline{\text{DO DSB}}$	23
74	$\overline{\text{PHOLD}}$	$\phi 2$	24
75	$\overline{\text{PRESET}}$	$\phi 1$	25
76	PSYNC	PHLDA	26
77	$\overline{\text{PWR}}$	PWAIT	27
78	PDBIN	PINTE	28
79	A0	A5	29
80	A1	A4	30
81	A2	A3	31
82	A6	A15	32
83	A7	A12	33
84	A8	A9	34
85	A13	DO1	35
86	A14	DO0	36
87	A11	A10	37
88	DO2	DO4	38
89	DO3	DO5	39
90	DO7	DO6	40
91	DI4	DI2	41
92	DI5	DI3	42
93	DI6	DI7	43
94	DI1	SM1	44
95	DI0	SOUT	45
96	SINTA	SINP	46
97	$\overline{\text{SWO}}$	SMEMR	47
98	SSTACK	SHLTA	48
99	POC	$\overline{\text{CLOCK}}$	49
100	GND	GND	50

Fig. 2-7. S-100 pinout designations.

Pin No.	Signal	Definition
11	VI7	Vectored interrupt 7.
12*	XRDY 2	External ready 2.
	$\overline{\text{NMI}}$	Nonmaskable interrupt.
	+12 V	Battery backup.
	READY 3	
13*	$\overline{\text{IRQ}}$	Interrupt request.
	CK3	Phase 3 shift clock.
	STDBY	Standby power.
	PAUSE	—
	—	Bank select 8.
14*	RDS BL/L	Phase.
	$\phi 1$	M6800 phase 1 clock
	CK1	Phase 1 shift clock.
	STDBY	Standby power.
	—	Bank select 9.
	OPREQ	
15*	$\phi 2$	M6800 Phase 2 clock.
	$\overline{\text{BWE}}$	Write enable ANDed with write.
	$\overline{\text{NMI}}$	Nonmaskable interrupt.
	—	Battery backup.
	—	Bank select 10/address 18.
	$\overline{\text{STOP CLK}}$	
16*	R/W	M6800 read/write.
	LDM	Load medium address byte.
	—	Bank select 11/address 16.
	INTAK	
17*	$\overline{\text{BDSEL}}$	Acknowledge signal from addressed memory board.
	$\overline{\text{NMI}}$	Nonmaskable interrupt.
	BGNT+	CPU bus grant.
	—	Bank select 12/address 17.
18	$\overline{\text{STAT DSB}}$	Status disable—A low on this line puts the eight status line buffers (SMEMR, SINP, SMI, SOUT, SHLTA, SSTACK, $\overline{\text{SWO}}$, and SINTA) into a high-impedance state.
19	$\overline{\text{C/C DSB}}$	Command/control disable—A low on this line puts the six command/control line buffers (PHLDA, PSYNC, PDBIN, PINTE, $\overline{\text{PWR}}$, and PWAIT) into a high-impedance state.
20	UNPROT	Unprotect—A positive pulse resets the protect flip-flop on the currently addressed board so that it can accept data. A positive pulse on PROT (pin 70) sets the flip-flop so that data cannot be written into the currently addressed memory board. Protect status ($\overline{\text{PS}}$, pin 69) indicates the flip-flop status; low for protect and high for unprotect.
21	SS	Single step—Used by front panel, a high disables input buffer while panel drives bidirectional data bus.
22	$\overline{\text{ADD DSB}}$	Address disable—A low on this line puts the 16 address line buffers into a high-impedance state.
23	$\overline{\text{DO DSB}}$	Data out disable—A low on this line puts the eight data line buffers into a high-impedance state.
24	$\phi 2$	Phase 2 clock—The master timing signal for the bus.
25	$\phi 1$	Phase 1 clock.
26	PHLDA	Halt acknowledge — Processor command/control output signal which goes high in response to the hold signal. It indicates that the data and address buses will go to the high-impedance state

Pin No.	Signal	Definition
		and the processor (8080) will enter hold state after completion of current machine cycle. Note: ADD DSB and DO DSB must be pulled low to float the system bus.
27	PWAIT	Wait—Command/control signal out which, when high, acknowledges that processor is in wait state.
28	PINTE	Interrupt enable—Command/control signal out which indicates condition of interrupt enable flip-flop. When set (high), interrupts are accepted. Flip-flop is set by enable interrupt (EI) instruction. It is reset by disable interrupt (DI) instruction, when an interrupt is accepted, and by reset signal.
29	A5	Address bit 5.
30	A4	Address bit 4.
31	A3	Address bit 3.
32	A15	Address bit 15.
33	A12	Address bit 12.
34	A9	Address bit 9.
35	DO1	Data out bit 1.
36	DO0	Data out bit 0.
37	A10	Address bit 10.
38	DO4	Data out bit 4.
39	DO5	Data out bit 5.
40	DO6	Data out bit 6.
41	DI2	Data in bit 2.
42	DI3	Data in bit 3.
43	DI7	Data in bit 7.
44	SM1	SM1 status output signal which, when high, indicates that the current bus cycle is an op-code fetch.
45	SOUT	Status output signal which, when high, indicates that the address bus contains the address of an output device and the data bus will contain the output data when PWR is active.
46	SINP	In—Status output signal which, when high, indicates that the address bus contains the address of an input device and the input data should be placed on the data bus when PDBIN is active.
47	SMEMR	Memory read—Status output signal which, when high, indicates that the data bus will be used to read memory data.
48	SHLTA	Halt acknowledge—Status output signal which, when high, acknowledges that a halt instruction has been executed.
49	CLOCK	Phase 2 clock inverted.
50	GND	Signal and power ground.
51	+8 V	Same as pin 1.
52	−16 V	Negative unregulated voltage supply.
53	SSW DSB	Sense switch disable—A low disables the data input buffers so the input from the sense switches may be strobed onto the bidirectional data bus.
54	EXT CLR	External clear—A low clears i/o devices. Generated by front panel switch.
55*	RTC	Real time clock.
	GND	Ground.
56*	STSTB	Status strobe—A low indicates that status signals are on data bus.
	DMAGR7	DMA grant 7.
	Vcc	Standby.
57*	DIG1	Data input gate 1.
	DMAGR6	DMA grant 6.

Pin No.	Signal	Definition
58*	FRDY	Front panel ready—When low, it disables WRITE driver.
	DMAGR5	DMA grant 5.
59*	DMAGR4	DMA grant 4.
	CW	Conditioned memory write.
	—	Memory disable (PGANTOM).
	EXINH0	Bank select 0.
	INH	Shadow memory.
	Program Switch	To enable programmer.
60*	DMAGR3	DMA grant 3.
	ABX	NOT alternate bank X (A16).
	MBS	Memory bank select.
	—	Bank select 1.
	MONITOR	Enable user's monitor.
61*	A17	Address line 17.
	DMAGR2	DMA grant 2.
	KRDY	KIMSI ready.
	—	Bank select 2.
	GND	Ground.
62*	DMAGR1	DMA grant 1.
	CK4	Phase 4 shift clock.
	Kϕ2	KIMSI ϕ2 clock.
	—	Bank select 3.
	A18	Address line 18
63*	DMAGR0	DMA grant 0
	CK2	Phase 2 shift clock.
	KHOLD	KIMSI hold.
	—	Bank select 4.
	A19	Address line 19.
64*	DMA RCVD	DMA grant flag.
	BCE	Chip enable.
	RTC	Real time clock (60 Hz).
	—	Bank select.
	RDSBL/H	
65*	MREQ	Memory request.
	WRITE	Write enable.
	—	Bank select 6.
66*	RFSH	Refresh.
	LDH	Load high address byte.
	CMCLK	Variable video clock.
	—	Bank select 7.
	8/16	
67*	PHANTOM	Phantom disable—Used to disable RAM and i/o addressing.
	NMI	Nonmaskable interrupt.
	RFSHDSBL	Refresh disable.
	MDSBL	Memory disable.
	BBSY+	Phantom.
	RFSH	Refresh.
	SCLK	Clock (video sample).
	A19	Address line 19.
68	MWRITE	Memory write—A high indicates that the current data on the data out bus is to be written into the memory location currently on the address bus.

Pin No.	Signal	Definition
69	$\overline{\text{PS}}$	Protect status—See pin 20.
70	PROT	Protect—See pin 20.
71	RUN	Run—A high indicates that the run/stop flip-flop is set to RUN on the front panel.
72	$\overline{\text{PRDY}}$	Ready—See pin 3.
73	$\overline{\text{PINT}}$	Interrupt request—A low causes the processor to recognize an interrupt request at the end of the current instruction or while halted. If the CPU is in the hold state or if the interrupt enable flip-flop is reset, it will not honor the request.
74	$\overline{\text{PHOLD}}$	Hold—A low requests the processor to enter the hold state. It allows an external device to gain control of the address and data buses as soon as the current machine cycle is completed.
75	$\overline{\text{PRESET}}$	Reset—A low causes the contents of the program counter to be cleared and the instruction register is set to 0.
76	PSYNC	Sync—The command/control signal out which, when high, identifies the beginning of a machine cycle.
77	$\overline{\text{PWR}}$	Write—The command/control signal out which, when low, signifies the presence of valid data on the data out bus for memory write or i/o output.
78	PDBIN	Data bus in—The command/control signal out which, when high, requests data on the DI bus from the currently addressed memory or i/o.
79	A0	Address bit 0.
80	A1	Address bit 1.
81	A2	Address bit 2.
82	A6	Address bit 6.
83	A7	Address bit 7.
84	A8	Address bit 8.
85	A13	Address bit 13.
86	A14	Address bit 14.
87	A11	Address bit 11.
88	DO2	Data out bit 2.
89	DO3	Data out bit 3.
90	DO7	Data out bit 7.
91	DI4	Data in bit 4.
92	DI5	Data in bit 5.
93	DI6	Data in bit 6.
94	DI1	Data in bit 1.
95	DI0	Data in bit 0.
96	SINTA	Interrupt acknowledge—The status output signal which, when high, identifies the instruction fetch cycle(s) that immediately follow an accepted interrupt request presented on $\overline{\text{PINT}}$.
97	$\overline{\text{SWO}}$	Write/output—The status output signal identifying a bus cycle which, when low, transfers data from processor to memory or i/o port.
98	SSTACK	Stack—Status output signal which indicates, when high, that the address bus holds the pushdown stack address from the stack pointer and that a stack operation will occur on the current cycle.
99	POC	Power on clear—Generated by $\overline{\text{PRESET}}$ or power on. Used to reset CPU and i/o devices.
100	GND	Signal and power ground.

DISCUSSION

As can be seen from the signal definitions, the S-100 bus pinout is generally defined except for those pins marked with an asterisk: pins 12–17 and 55–67. In the signal definitions column, some of the known uses for these lines have been listed; some new ones, however, have probably been omitted. Some systems that are not 8080-based systems do not generate all of the signals listed. If a board requires one or more of these missing signals, then a real problem exists which may require extensive modifications.

To use the S-100 bus effectively, care must be taken in reading the specifications of all boards used or contemplated being used. All boards must have the proper lines on the proper pin or capability for jumpering to the proper pin. All signals required by any of these boards must be generated somewhere in the system. Information on systems and boards may be obtained from manufacturers, computer stores, computer clubs, and other hobbyists.

CHAPTER 3

The IEEE S-100 Bus

This chapter discusses the proposed IEEE standard S-100 bus (IEEE Task 696 1/D2). The proposal upgrades the bus to 16-bit operation with 24 address lines (16 megabytes). By configuring the bus for both 8- and 16-bit data transfers, system flexibility is increased by allowing both 8- and 16-bit masters and i/o devices. Several other lines have been added, some have been deleted, and some have been clarified. Chapter 2 discusses the S-100 bus as it has existed from its early days.

The IEEE S-100 specifies a maximum bus transmission length of 25 inches, a maximum of 22 devices (boards) on the bus, and a maximum data rate of 6 MHz.

Figs. 3-1 and 3-2 show a mainframe and 16-bit Z-8000 CPU board respectively, both configured to the IEEE S-100 bus. The proposed S-100 pinout designations are shown in Fig. 3-3. It should be noted that the active-low lines are designated by asterisks rather than bars over the names. This was done to avoid confusion with logical negation.

There are three types of signals (M, S, and B) on the S-100 bus. Signals designated by (M) are those which a bus master must generate. The bus master currently controlling the bus has the responsibility for faithfully generating all type M signals during the duration of its control. Signals designated (S) are those which bus slaves generate. A slave need only generate that subset of type S signals nec-

Courtesy Intersystems

Fig. 3-1. Intersystems S-100 mainframe.

essary to communicate with bus masters able to address it. Signals designated by (B) are by default all those which are not type M or S. This is not to say that some bus master may not generate a type B signal. Rather, a type B signal is one that (a) not all bus masters are required to generate and (b) not any bus slave is required to generate.

Courtesy Intersystems

Fig. 3-2. Intersystems Z-8000 IEEE S-100 board.

Pin	Signal	Signal	Pin
51	+8V	+8V	1
52	−16V	+16V	2
53	GND	XRDY	3
54	SLAVE CLR*	VI0*	4
55	DMA0*	VI1*	5
56	DMA1*	VI2*	6
57	DMA2*	VI3*	7
58	sXTRQ*	VI4*	8
59	A19	VI5*	9
60	SIXTN*	VI6*	10
61	A20	VI7*	11
62	A21	NMI*	12
63	A22	PWRFAIL*	13
64	A23	DMA3*	14
65	NDEF	A18	15
66	NDEF	A16	16
67	PHANTOM*	A17	17
68	MWRT	SDSB*	18
69	RFU	CDSB*	19
70	GND	GND	20
71	NDEF	RFU	21
72	RDY	ADSB*	22
73	INT*	DODSB*	23
74	HOLD*	ϕ	24
75	RESET	pSTVAL*	25
76	pSYNC	pHLDA	26
77	pWR*	RFU	27
78	pDBIN	RFU	28
79	A0	A5	29
80	A1	A4	30
81	A2	A3	31
82	A6	A15	32
83	A7	A12	33
84	A8	A9	34
85	A13	DO1/DATA1	35
86	A14	DO0/DATA0	36
87	A11	A10	37
88	DO2/DATA2	DO4/DATA4	38
89	DO3/DATA3	DO5/DATA5	39
90	DO7/DATA7	DO6/DATA6	40
91	DI4/DATA12	DI2/DATA10	41
92	DI5/DATA13	DI3/DATA11	42
93	DI6/DATA14	DI7/DATA15	43
94	DI1/DATA9	sM1	44
95	DI0/DATA8	sOUT	45
96	sINTA	sINP	46
97	sWO*	sMEMR	47
98	ERROR*	sHLTA	48
99	POC*	CLOCK	49
100	GND	GND	50

Fig. 3-3. IEEE S-100 pinout designations.

S-100 BUS SIGNAL DEFINITIONS

Pin No.	Signal	Definition
1	+8V (B)	+8 volts (B)—Unregulated input to +5V regulators.
2	+16V (B)	+16 volts (B)—Positive unregulated voltage supply.
3	XRDY (S)	Ready—The ready lines are used by bus slaves to synchronize bus masters to the response speed of the slave. Thus bus cycles are suspended and wait states inserted until both ready lines are asserted. The RDY line is the general ready line for bus slaves. It is specified as an open-collector line. The XRDY line is a special ready line commonly used by front-panel devices to stop and single-step bus masters. As it is not specified as open collector, it should not be used by other bus slaves.
4	VI0* (S)	Vectored interrupt line 0—Open collector. Highest priority. See discussion for further explanation of VI lines.
5	VI1* (S)	Vectored interrupt line 1. Open collector.
6	VI2* (S)	Vectored interrupt line 2. Open collector.
7	VI3* (S)	Vectored interrupt line 3. Open collector.
8	VI4* (S)	Vectored interrupt line 4. Open collector.
9	VI5* (S)	Vectored interrupt line 5. Open collector.
10	VI6* (S)	Vectored interrupt line 6. Open collector.
11	VI7* (S)	Vectored interrupt line 7. Open collector. Lowest priority.
12	NMI* (S)	Nonmaskable interrupt request—The nonmaskable interrupt line is not maskable by the bus master. It asserts an interrupt on the negative-going edge and need not assert an interrupt acknowledge. Open collector.
13	PWRFAIL*(B)	Power fail—The power fail line indicating impending power failure. It is pulled low when loss of power is sensed and remains true until power is restored and POC* is true. It must be asserted at least 50 ms before local regulators drift out of specifications. It may be used with NMI* for orderly shutdown routine.
14	DMA3* (M)	Direct memory access 3—One of the four (DMA0*, DMA1*, DMA2*, and DMA3*) lines used to establish priority among temporary bus masters when two or more are requesting the bus simultaneously. Open collector.
15	A18 (M)	Extended address bit 18.
16	A16 (M)	Extended address bit 16.
17	A17 (M)	Extended address bit 17.
18	SDSB* (M)	Status disable—Disables (tri-states) the eight status lines (sMEMR, sM1, sINP, sOUT, sWO*, sINTA, sHLTA, sXTRQ*) from the permanent bus master when asserted. Open collector.
20	GND (B)	Ground line—Common to line 100.
21	NDEF	Not to be defined. Manufacturer must specify any use in detail.
22	ADSB* (M)	Address disable—Disables (tri-states) the 24 address lines (A0–A23) from the permanent bus master when asserted. Open collector.
23	DODSB* (M)	Data out disable—Disables (tri-states) the eight data out lines (DO0–DO7) from the permanent bus master when asserted. Open collector. It would also have to disable the other eight data lines for a 16-bit permanent master.

Pin No.	Signal	Definition
24	ϕ (B)	ϕ (B)—The master timing signal for the bus. Usually generated on the permanent master board.
25	pSTVAL* (M)	Status valid strobe—Decoded in conjunction with pSYNC indicates that stable address and status may be sampled from the bus in the current cycle.
26	pHLDA (M)	Hold acknowledge—Signal that indicates to the highest-priority temporary master that the permanent master is relinquishing control of the bus. The permanent master's lines will be tri-stated during the next bus cycle.
27	RFU	Reserved for future use.
28	RFU	Reserved for future use.
29	A5	Address bit 5.
30	A4	Address bit 4.
31	A3	Address bit 3.
32	A15	Address bit 15.
33	A12	Address bit 12.
34	A9	Address bit 9.
35	DO1 (M)/ DATA1 (M/S)	Data out bit 1/Bidirectional data bit 1.
36	DO0 (M)/ DATA0 (M/S)	Data out bit 0/Bidirectional data bit 0.
37	A10 (M)	Address bit 10.
38	DO4 (M)/ DATA4 (M/S)	Data out bit 4/Bidirectional data bit 4.
39	DO5 (M)/ DATA5 (M/S)	Data out bit 5/Bidirectional data bit 5.
40	DO6 (M)/ DATA6 (M/S)	Data out bit 6/Bidirectional data bit 6.
41	DI2 (M)/ DATA10 (M/S)	Data in bit 2/Bidirectional data bit 10.
42	DI3 (M)/ DATA11 (M/S)	Data in bit 3/Bidirectional data bit 11.
43	DI7 (M)/ DATA15 (M/S)	Data in bit 7/Bidirectional data bit 15.
44	sM1 (M)	M1 cycle—The status signal which indicates that the current cycle is an op-code fetch.
45	sOUT (M)	Output—The status signal identifying the data transfer bus cycle to an output device.
46	sINP (M)	Input—The status signal identifying the data transfer bus cycle from an input device.
47	sMEMR (M)	Memory read—The status signal identifying bus cycles which transfer data from memory to a bus master, which are not interrupt-acknowledge instruction-fetch cycle(s).
48	sHLTA (M)	Halt acknowledge—The status signal which acknowledges that a HLT instruction has been executed.
49	CLOCK (B)	Clock—2 MHz (0.5 percent) 40–60-percent duty cycle. Not required to be synchronous with any other bus signal.
50	GND (B)	Ground—The ground line common to pin 100.
51	+8V (B)	+8 volts (B)—Unregulated input to +5V regulators.
52	−16V (B)	−16 volts (B)—Negative unregulated voltage supply.
53	GND (B)	Ground—Ground line common to pin 100.
54	SLAVE CLR* (B)	Slave clear—Reset signal to reset bus slaves. Must be as-

Pin No.	Signal	Definition
		serted with POC* and may also be generated externally. Open collector.
55	DMA0* (M)	Direct memory access 0—See pin 14.
56	DMA1* (M)	Direct memory access 1—See pin 14.
57	DMA2* (M)	Direct memory access 2—See pin 14.
58	sXTRQ* (M)	Sixteen request—Status output of bus master which indicates a request for a 16-bit data transfer.
59	A19 (M)	Extended address bit 19.
60	SIXTN* (S)	Sixteen acknowledge—A slave responds to sXTRQ* indicating if true that the requested 16-bit data transfer is possible. Open collector.
61	A20 (M)	Extended address bit 20.
62	A21 (M)	Extended address bit 21.
63	A22 (M)	Extended address bit 22.
64	A23 (M)	Extended address bit 23.
65	NDEF	Not to be defined. See pin 21.
66	NDEF	Not to be defined. See pin 21.
67	PHANTOM* (M/S)	Phantom—Allows overlaying bus slaves at a common address location. When asserted phantom bus slaves are enabled and normal bus slaves are disabled. Primarily used for bootstrapping systems without front panels. Open collector.
68	MWRT (B)	Memory write—Memory write strobe defined as (pWR) · (−sOUT) a true pWR* logically ANDed with the logical negation of sOUT.
69	RFU	Reserved for future use.
70	GND (B)	Ground—Ground line common to pin 100.
71	RFU	Reserved for future use.
72	RDY (S)	Ready—Ready line; see pin 3. Open collector.
73	INT* (S)	Interrupt—Used to request service from permanent bus master but can usually be masked by software operation. Often used in conjunction with the vectored interrupt lines to establish priorities.
74	HOLD* (M)	Hold request line—Used by temporary bus masters to request control of the bus from the permanent master. May be masked by the permanent master. Open collector.
75	RESET* (B)	Reset—The signal to reset bus master devices. Must be active with POC* and may be generated externally. Open collector.
76	pSYNC (M)	Sync—The control signal identifying BS_1. Bus state 1 (BS_1) is the initial bus state in a bus cycle. During this time the address and status lines are settling.
77	pWR* (M)	Write—A generalized write strobe that writes data from the data bus into an addressed slave.
78	pDBIN (M)	Data bus in—A generalized read strobe that gates data from an addressed slave onto the data bus.
79	A0 (M)	Address bit 0. Least significant.
80	A1 (M)	Address bit 1.
81	A2 (M)	Address bit 2.
82	A6 (M)	Address bit 6.
83	A7 (M)	Address bit 7.
84	A8 (M)	Address bit 8.
85	A13 (M)	Address bit 13.
86	A14 (M)	Address bit 14.
87	A11 (M)	Address bit 11.

Pin No.	Signal	Definition
88	D02 (M)/ DATA2 (M/S)	Data out bit 2/Bidirectional data bit 2.
89	D03 (M)/ DATA3 (M/S)	Data out bit 3/Bidirectional data bit 3.
90	D07 (M)/ DATA7 (M/S)	Data out bit 7/Bidirectional data bit 7.
91	DI4 (S)/ DATA12 (M/S)	Data in bit 4/Bidirectional data bit 12.
92	DI5 (S)/ DATA13 (M/S)	Data in bit 5/Bidirectional data bit 13.
93	DI6 (S)/ DATA14 (M/S)	Data in bit 6/Bidirectional data bit 14.
94	DI1 (S)/ DATA9 (M/S)	Data in bit 1/Bidirectional data bit 9.
95	DI0 (S)/ DATA8 (M/S)	Data in bit 0/Bidirectional data bit 8.
96	sINTA (M)	Interrupt Acknowledge—The status signal identifying the bus input cycle(s) that may allow an accepted interrupt request on INT*. Some type of vector fetch would normally be done during sINTA.
97	sWO* (M)	Write output—The status signal identifying a bus cycle which transfers data from a bus master to a slave.
98	ERROR* (S)	Error—Generalized error line that is asserted when an error of some sort (for example, parity, write to protected memory, inability to accommodate 8-bit slaves, etc.) is occurring in the current bus cycle. The ERROR* line should be implemented as a trap. All relevant information about the cycle should be latched on the falling edge of ERROR*. NMI* may be used to enter an error finding/correction routine.
99	POC* (B)	Power-on clear—Clears all bus devices at power-on.
100	GND (B)	Ground—System ground (power and signal).

DISCUSSION

Only the principal changes in the proposed S-100 from the previous chapter are discussed. The other lines retain the same information as discussed in Chapter 2.

All 100 lines are specified precisely except for the three NDEFs (21, 65, 66) and the four RFUs (27, 28, 69, 71). A bus user can specify the NDEFs as he or she wishes but the RFUs are reserved for future use and should not be used.

There have been three ground lines added to make signal grounds more available on boards. These are lines 20, 53, and 70.

The address bus has been expanded to 24 bits by adding A16–A23 (lines 16, 17, 15, 29, 61, 62, 63, and 64 respectively). The old data-in and data-out lines have been specified so as to keep the same designation for 8-bit data transfers, but they become a 16-bit directional bus for 16-bit transfers.

The two new signals, sixteen request (sXTRQ*, line 58) and sixteen acknowledge (SIXTN*, line 60), allow a 16-bit bus master to transfer 8- or 16-bit data. Even with a 16-bit master, 8-bit data transfers would occur when doing 8-bit op-code fetches and 8-bit data transfers to/from memory and other slave devices. One type of operation would be a 16-bit data transfer. For this the master would assert sXTRQ* and the slave would assert SIXTN*, indicating that the transfer was possible. If an 8-bit slave was addressed, SIXTN* would not be asserted and the master would have to do the transfer in two bytes or, alternately, an ERROR* signal could be generated.

If memory, or other slaves, are organized in an 8/16-bit configuration, as two banks (high byte and low byte), then any byte or word transfer can be successfully accomplished using 8- or 16-bit masters. If a 16-bit master is used, for maximum use of memory, the master must be able to address any byte of memory and also address 16-bit words on odd or even bytes. This is because operations may involve either a one- or two-byte op code and either one or two bytes of data. To simplify the logic circuitry some systems may use simple logic and only address 16-bit words on even bytes. On 16-bit operations, only A1–A23 are used. On 8-bit operations, A1–A23 address memory and A0 selects either the high byte (A0=0) or the low byte (A0=7).

The proposed S-100 can accommodate one permanent bus master (possibly a dummy master) and up to 16 temporary bus masters. A dummy master is a device that conducts no bus cycles but only supplies an arbitration interval so that the DMA control bus may settle. The temporary masters have a priority established by the four direct memory access lines (DMA0*, DMA1*, DMA2*, and DMA3*). These lines are used only by the temporary bus masters. A temporary master desiring the bus will assert the HOLD* line to the permanent bus master and the proper DMA lines according to its priority. If the 4 bits of the DMA lines match the temporary master's priority, then it receives control of the bus. If the 4 bits of the DMA lines do not match the temporary master's priority, then some other master or masters are also requesting the bus. Each temporary master which does not match the highest-order bit DMA3* will release its control of the remaining DMA lines and so on down through DMA2*, DMA1*, and DMA0* until only one of the sixteen possible codes remains. That device receives control of the bus after pHLDA is asserted by the permanent master. Then the other requesting temporary masters will release the HOLD* line. The temporary master must also assert the disable lines (ADSB*, SDSB*, DODSB*, and CDSB*) to remove the permanent master from the bus. When the temporary master is finished it releases the HOLD* line, and the permanent bus master will output pHLDA false (low). This will cause any requesting temporary bus masters to reassert HOLD* and their priorities on the DMA lines. This continues until all DMA requests are satisfied. Temporary bus masters are not subject to DMA signals. This prevents nested DMA operations.

The vectored interrupt system on the IEEE S-100 is designed to utilize an interrupt controller in the system. The controller would accept the eight vectored interrupt (VI0–VI7) signals, which could then be masked and rotated and an interrupt signal asserted on the INT* line to an interruptable master. VI0 is designated as the highest priority, but the masking and/or rotation of the interrupt controller can alter that. When a master asserts interrupt acknowledge (sINTA), the interrupt controller must put the proper vectoring information on the data bus. The interrupt controller is not addressed; therefore, multiple interrupt controllers would have to be daisy chained.

The external clear ($\overline{\text{EXT CLR}}$, line 54) line has been redesignated SLAVE CLR* and is used to reset all bus slaves.

A nonmaskable interrupt (NMI*, line 12) has been added, as are also a line to indicate impending power failure (PWRFAIL*, line 13) and a line (ERROR*, line 98) to indicate when some sort of error is occurring. Situations creating an ERROR* true might include parity, 16-bit read/write to an 8-bit slave, write to protected memory, and so on.

A pSTVAL* line which indicates that the address and status lines are stable was added. Line 49 was changed from $\overline{\text{CLOCK}}$, which was normally $\phi 2$ inverted, to CLOCK, which is a 2-MHz signal not synchronized to any other bus signal. CLOCK is intended to be used by timers, counters, and the like.

Among the signals which have been deleted by the proposed S-100 are the 8080-derived PINTE, PWAIT, SSTACK, 01, and the status strobe $\overline{\text{STSTB}}$ used in 8080 systems. The three signals used to protect memory boards (UNPROT, PROT, and $\overline{\text{PS}}$) have also been deleted. Other deleted signals are XRDY2, SS, RTC, and RUN.

The proposed S-100 is designed to allow both 8- and 16-bit masters and slaves to operate in a system. It is expected that newly designed boards will plug into the bus and operate properly. But there are thousands of older boards "out there" that may require some modification for use with the new proposed standard. The not-to-be-defined lines give some flexibility to make changes in boards in order to meet the proposed S-100 standard. Note that it is possible that a system may have both 8- and 16-bit bus masters and slaves.

CHAPTER 4

The Benton Harbor Bus

The Benton Harbor bus was developed by Heath Company for use on their 8080-based H8 system (shown in Fig. 4-1). Heath Company now offers the H8 system, which includes the H8 (chassis, power supply, motherboard, control board, and CPU board), the H8-1 (8K × 8 memory board with 4K of memory), the H8-2 (parallel interface board), the H8-3 (4K × 8 memory for the H8-1), and the H8-5 (serial i/o and cassette interface). Heath Company also offers the H10 papertape reader/punch and the H9 video terminal (both shown in Fig. 4-1), and the H8 floppy disk (shown in Fig. 4-2). Heath Company plans to offer other boards and accessories. Other companies have offered boards for the Heath H-8 system.

Courtesy Heath Co.

Fig. 4-1. The H8 system with H9 video terminal and H10 papertape reader/punch.

The approximately 5½- by 14⅜-inch (13.9- by 36.5-cm) motherboard (Fig. 4-3) is mounted vertically and has slots (P1–P10) for ten other boards to be inserted at an unusual angle. Slot P1 is reserved for the H8 control board, P2 is reserved for the H8 CPU board, and slot P10 is reserved for an expansion connector. That leaves seven slots for expansion within the H8 mainframe.

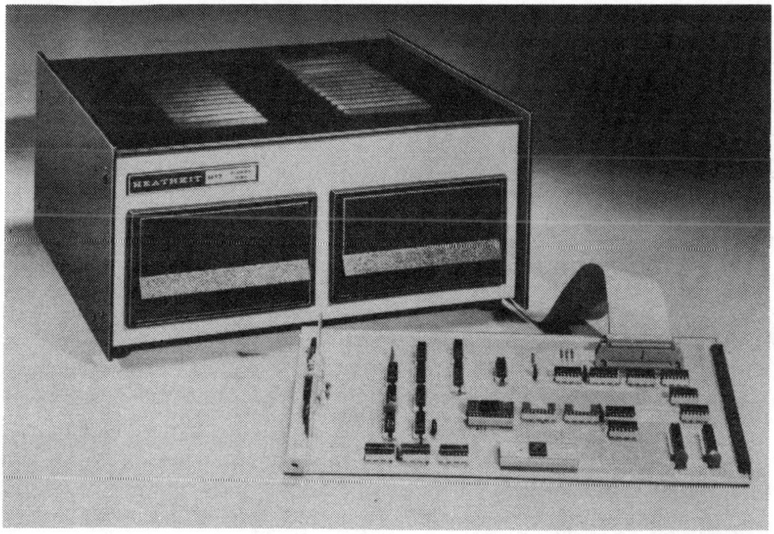

Courtesy Heath Co.

Fig. 4-2. The H8 floppy disk system.

A board for the Benton Harbor bus is approximately 6 by 12 inches (15.2 by 30.4 cm) with two 25-pin connectors on one end. The connectors slide onto the pins on the motherboard. See Figs. 4-4 and 4-5 for typical boards. A bracket on the opposite end from the connectors supports the boards when they are inserted into the H8 chassis.

BENTON HARBOR BUS SIGNAL DEFINITIONS

The Benton Harbor bus pinout designations are shown in Fig. 4-6.

Pin No.	Signal	Definitions
10–17	$\overline{D0}$–$\overline{D7}$	Data bus—The tri-state data bus provides bidirectional communication between the CPU, memory, and i/o devices for instructions and data transfers. D0 is least significant bit. Negative true.
23	MEMW	Memory write—A high strobes data on the data bus to the currently addressed memory or memory-mapped i/o devices.

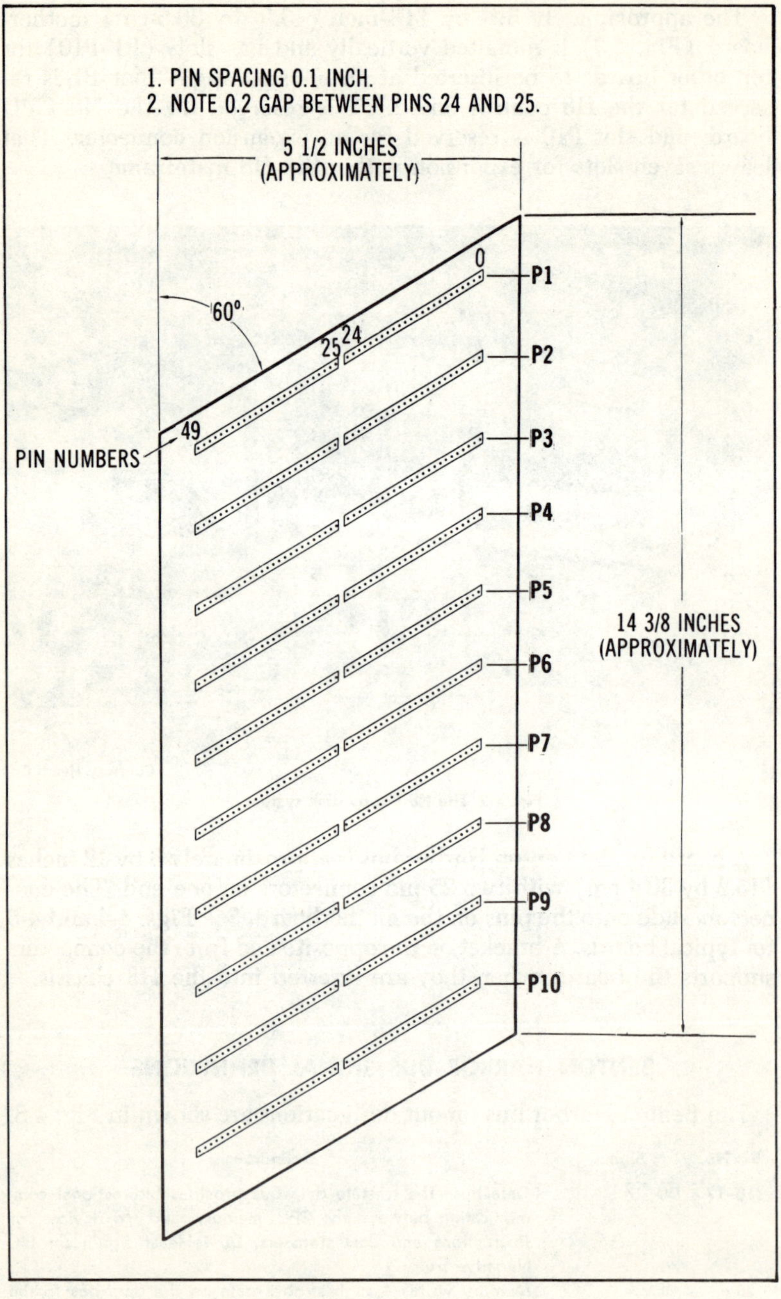

Fig. 4-3. Diagram of the Benton Harbor bus motherboard used in the H8.

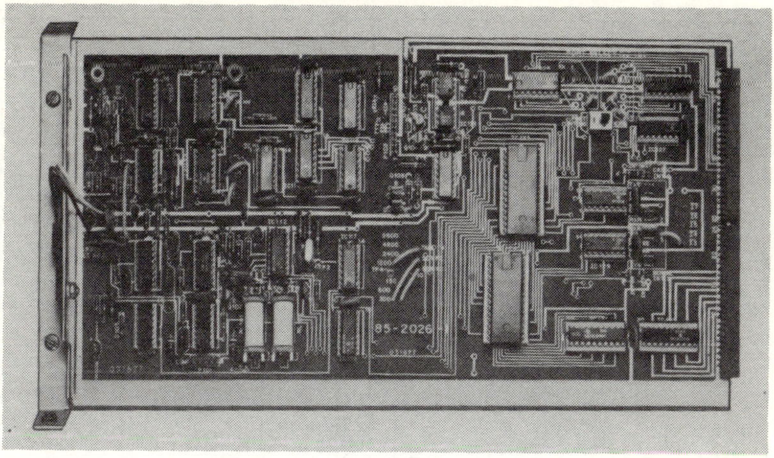

Courtesy Heath Co.

Fig. 4-4. H8-5 serial i/o and cassette interface board.

Pin No.	Signal	Definitions
21	I/OW	I/o write—A high strobes data on the data bus to the output device currently addressed by the lower 8 bits of the address bus.
28	MEMR	Memory read—A high strobes data from the addressed memory or memory-mapped i/o device onto the data bus and into the CPU.
26	I/OR	I/o read—A high strobes data from the addressed i/o device onto the data bus and into the CPU.

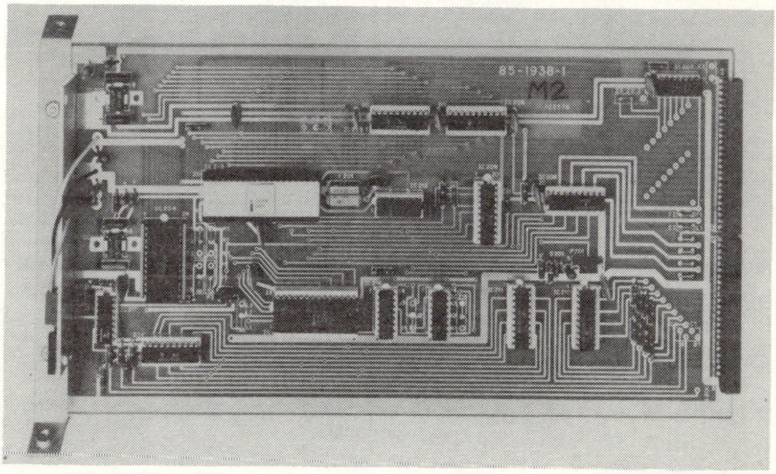

Courtesy Heath Co.

Fig. 4-5. H8 CPU board.

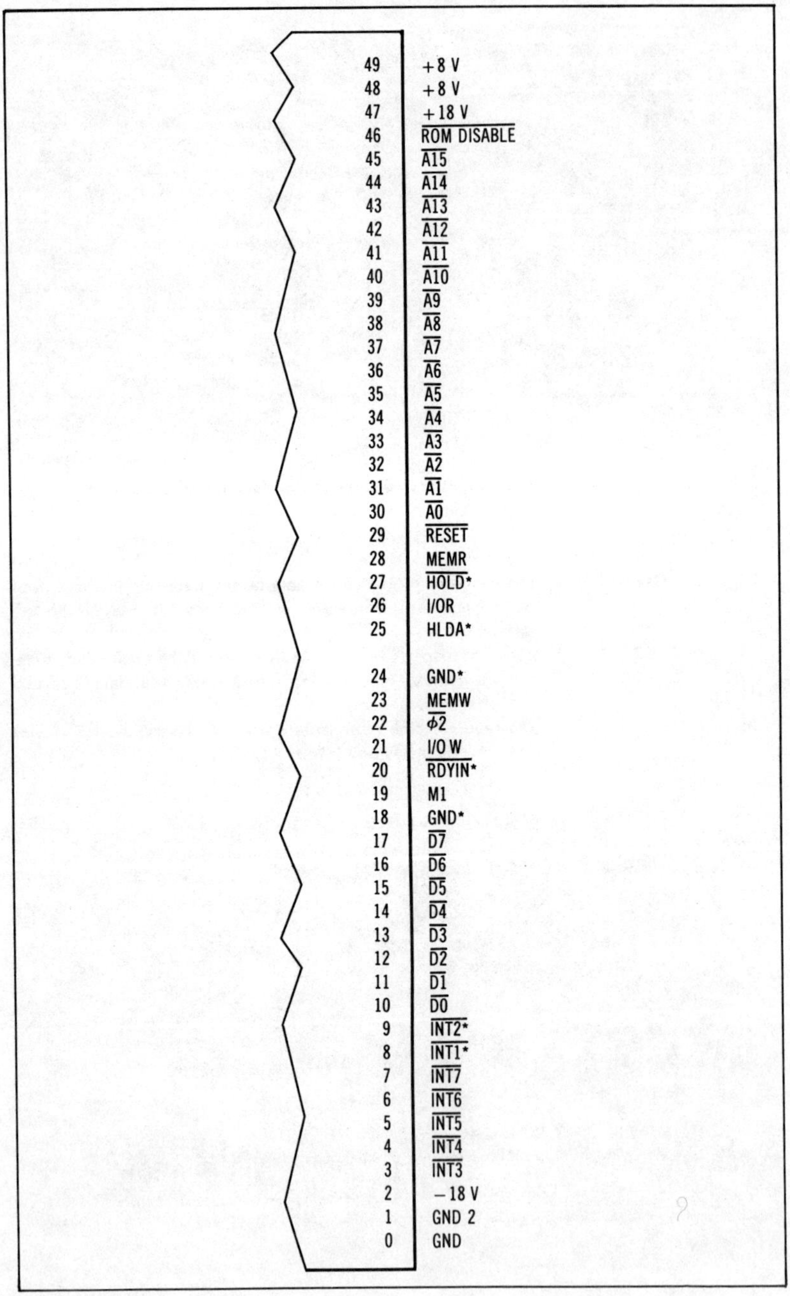

Fig. 4-6. Benton Harbor bus pinout designations.

Pin No.	Signal	Definitions
30–15	$\overline{A0}$–$\overline{A15}$	Address bus—Sixteen fully buffered lines that can select any one of 65,536 memory locations or 256 i/o ports. Negative true. A0 is least significant bit.
19	M1	Machine cycle one—A high signifies that the machine cycle being executed is an instruction fetch.
20	$\overline{\text{RDYIN}}$*	Ready in—A low causes the processor to enter a wait state until pulled high again. Used by slow memory.
27	$\overline{\text{HOLD}}$*	Hold—A low causes the CPU to enter a hold state at the end of the current machine cycle. The address and data buses will be tri-stated, allowing a DMA device to control these buses.
25	HLDA*	Hold acknowledge—A high appears in response to the hold signal and indicates that the data and address buses will be tri-stated.
46	$\overline{\text{ROM DISABLE}}$	When the X1, X2 jumper on the H8 CPU board is installed, a low on this signal disables the on-board ROM, allowing a test board or other ROM board to use the monitor address space.
3–7	$\overline{\text{INT3}}$–$\overline{\text{INT7}}$	Interrupts—A low on any combination of these lines activates the interrupt request of the CPU.
8, 9	$\overline{\text{INT1}}$, $\overline{\text{INT2}}$	Not dedicated—The H8 CPU board has jumper pads available to jumper INT1 and INT2 to these bus lines in order to share the front-panel interrupts, if desired. Otherwise, these lines may be used by the customer. With jumper installed, a low will activate an interrupt request of the CPU.
29	$\overline{\text{RESET}}$	Reset—Goes low in response to a manual reset or CPU power on. Used to put devices on bus in a known state for an orderly start up.
22	$\overline{\phi 2}$	Phase 2 clock—This buffered inverted signal is used to synchronize system devices connected to the bus.
0, 1	GND	Ground—Signal and power ground for bus.
18, 24	CPU GND	CPU ground—These lines are grounded on the H8 CPU board but can be floated for other use by removing one jumper.
2	−18 V	−18 V—Unregulated negative supply voltage.
47	+18 V	+18 V—Unregulated positive supply voltage.
48, 49	+8 V	+8 V—Unregulated positive supply voltage.

* Heath Company reserves the right to change these pin designations.

DISCUSSION

Heath Company in designing the Benton Harbor bus has stuck to the basics. Only the address bus, data bus, and generalized control lines are included. Specialized control lines are omitted so that the bus does not need to change with future alterations. A convenient design consideration on the bus is the positioning of the data bus, address bus, and interrupt lines so that the last digit of their respective number coincides with the last digit of the bus line (for example, $\overline{A6}$ is on line 36).

CHAPTER 5

The TRS-80 Model I Bus

The TRS-80 Model I, shown in Fig. 5-1, is Radio Shack's very popular microcomputer system. The keyboard enclosure, shown removed in Fig. 5-2, houses the CPU, RAM, executive and BASIC ROMs, the keyboard, video and cassette interfaces, and the power supply.

Courtesy Radio Shack Corp.

Fig. 5-1. The TRS-80 Model I microcomputer system.

Courtesy Radio Shack Corp.

Fig. 5-2. The keyboard unit of the TRS-80 Model I system.

Floppy disk units and printers (not shown) are available to expand the system.

Expansion of the Z-80 based system is accomplished via the 40-pin (0.1-inch, or 0.25-cm, spacing) edge connector shown in the upper right-hand corner of Fig. 5-3. No positive indexing is provided. All

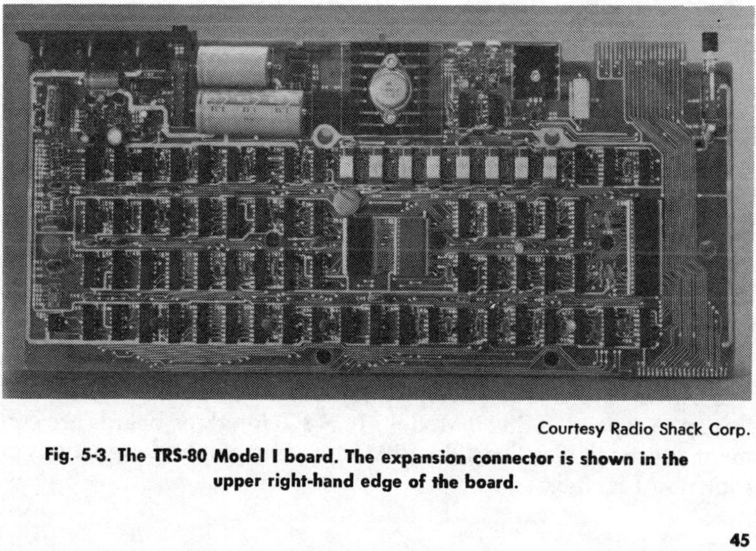

Courtesy Radio Shack Corp.

Fig. 5-3. The TRS-80 Model I board. The expansion connector is shown in the upper right-hand edge of the board.

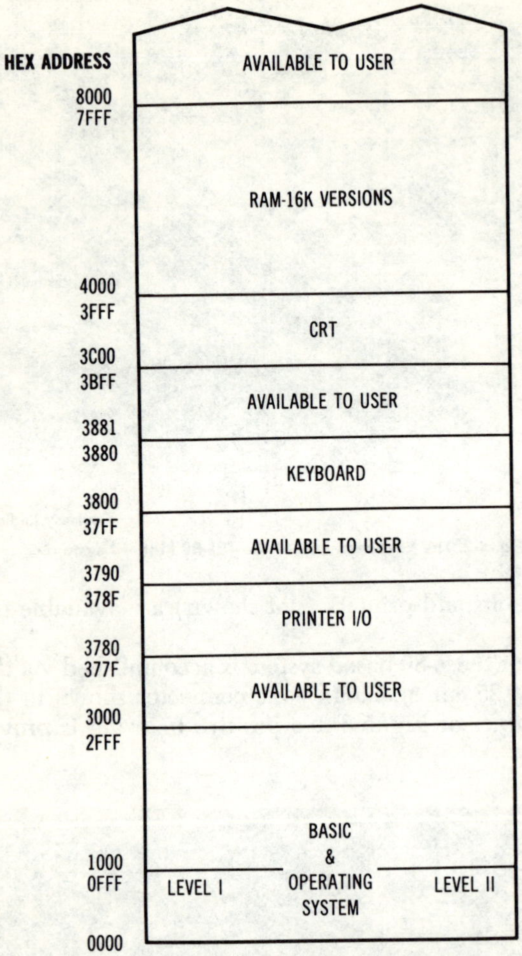

Courtesy Radio Shack Corp.

Fig. 5-4. Memory map for the TRS-80 Model I.

lines, except as noted, are buffered. Use of dynamic RAM is simplified by address multiplex signals that are brought to the connector. The memory space above hex 8000 is available for expansion. This is shown in the memory map given in Fig. 5-4.

Because its bus is relatively straightforward, the TRS-80 Model I is easy to expand. TRS-80 Model I to S-100 interface boards are commercially available. The TRS-80 Model I bus pinout designations are shown in Fig. 5-5.

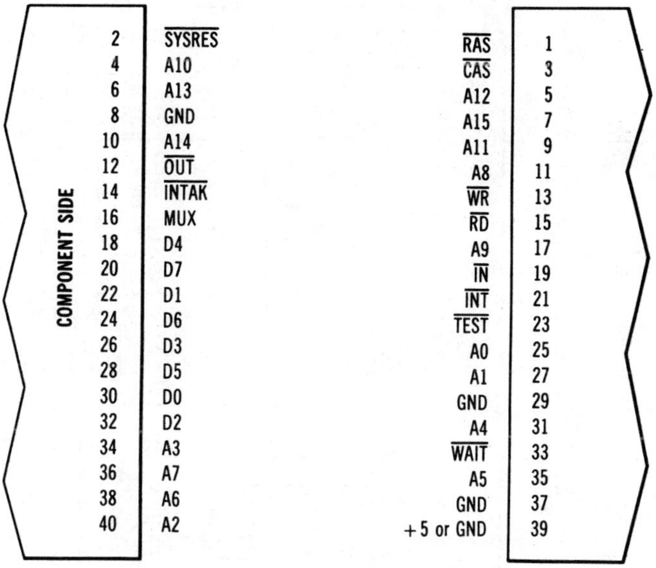

Fig. 5-5. Pinout designations for the TRS-80 Model I bus.

TRS-80 MODEL I BUS SIGNAL DEFINITIONS

Pin No.	Signal	Definition
18, 20, 22, 24, 26, 28, 30, 32	D0–D7	Data bus—The bidirectional data bus carries data between the processor and peripherals.
12, 13, 15, 19	$\overline{\text{RD}}$, $\overline{\text{WR}}$, $\overline{\text{IN}}$, $\overline{\text{OUT}}$	Read, write, input, output—The bus read, write, input, and output lines control the memory and i/o operations. They are derived from the read $\overline{\text{(RD)}}$, write $\overline{\text{(WR)}}$, memory request $\overline{\text{(MREQ)}}$, and i/o request $\overline{\text{(IORQ)}}$ outputs of the Z-80 as shown in Fig. 5-6. The processor's read and write lines are active-low outputs used to determine the direction of flow on the data bus. The $\overline{\text{MREQ}}$ line is an active-low output used to indicate a memory-read or -write operation and the $\overline{\text{IORQ}}$ line is an active-low output used to indicate an input or output operation. The bus $\overline{\text{RD}}$ line is the ORed combination of the processor's read and $\overline{\text{MREQ}}$ lines and goes low to indicate a memory-read operation. The bus $\overline{\text{WR}}$ line is the ORed combination of the processor's write and $\overline{\text{MREQ}}$ lines and goes low to indicate a memory read. The bus IN line is an active-low output that indicates an input operation, derived by ORing the $\overline{\text{RD}}$ and $\overline{\text{IORQ}}$ of the processor. The bus $\overline{\text{OUT}}$ line is an active-low output that indicates an output operation, derived by ORing the processor's $\overline{\text{WR}}$ and $\overline{\text{IORQ}}$.
1, 3, 16	$\overline{\text{RAS}}$, $\overline{\text{CAS}}$, MUX	Row address select, column address select, address multiplex—These lines are used in addressing industry standard

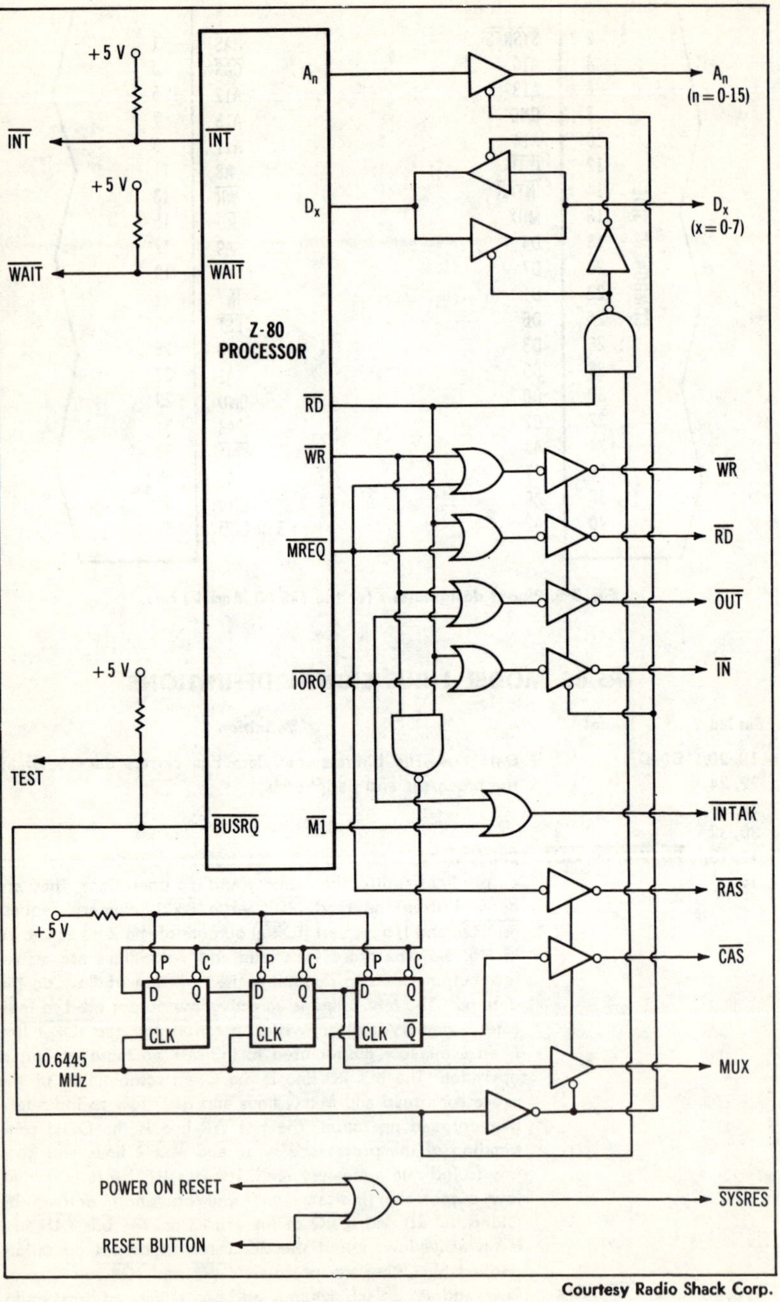

Courtesy Radio Shack Corp.

Fig. 5-6. Diagram showing a Z-80-to-bus interface.

Pin No.	Signal	Definition
		16-pin 4-kilobyte dynamic RAMs. Since only six address pins, A0–A5, are available on the RAM integrated circuit, the address must be fed to it in parts—multiplexed. Internally, the RAM consists of a 64-row by 64-column array of memory cells. Any cell can be located by specifying its row address and column address. Internal circuitry can latch the row address from A0–A5 on command, and then, on command, latch the column address from the same pins to form the complete address. External circuitry must accept the address from the processor's address bus, split it into row and column halves and, then, feed it sequentially to the RAM's address lines. A typical interface circuit is shown in Fig. 5-7. All buffers and the data bus circuitry have been omitted to simplify the sketch. Address lines A0–A11 feed the data multiplexers in pairs of row and column addresses (A0 is paired with A6, etc.). Only one input of each pair is passed to the RAM at one time. The signal on the multiplex input determines which group of address inputs is selected. At the beginning of each read or write cycle, the $\overline{MREQ}$ is pulled low to indicate a valid address. The $\overline{MREQ}$ is also used as the RAM row-address select signal ($\overline{RAS}$) for the RAM which latches the signals on A0–A5. These signals are from address bus lines A0–A5. When data is stable on D0–D7, the $\overline{RD}$ or $\overline{WR}$ line of the Z-80 is pulled low. The $\overline{RD}$ and $\overline{WR}$ are inverted and ORed to feed the flip-flop chain. The clear and FF1 data inputs are controlled by the OR gate. When either the $\overline{RD}$ or $\overline{WR}$ goes low, the clears are pulled high (inactive) and a logical 1 is loaded into flip-flop FF1 on the next 10.6445-MHz clock pulse. The 10.6445-MHz clock is obtained from the master system clock which also feeds the Z-80 microprocessor. The second clock pulse transfers the logical 1 to flip-flop FF2, where it appears on output Q2. The Q2 signal is the multiplex signal (MUX) that controls the data multiplexers. When the multiplex line goes high, the multiplexers change their outputs to reflect the A6–A11 inputs. The next clock pulse shifts the logical 1 to flip-flop FF3. Output $\overline{Q3}$ of FF3 is the RAM column-address select signal $\overline{CAS}$, which causes the RAM to latch the A6–A11 signals now appearing on its A0–A5 inputs. The RAM now has the complete address of the desired memory cell. All of this must occur in a short period since the RAM must make available, or be able to accept, data from the bus while $\overline{RD}$ or $\overline{WR}$ is low. When $\overline{RD}$ or $\overline{WR}$ goes high, the flip-flop clear lines are pulled low clearing them and resetting the multiplexers to reflect the Z-80 A0–A5 lines.
2	$\overline{SYSRES}$	System reset—The system reset line is the output of an internal reset circuit. It is pulled low during power-up or when the RESET button is pushed and it can be used to reset peripherals.
4–7, 9–11, 17, 25,	A0–A15	Address bus—The address bus lines output the location of the byte desired by the processor.

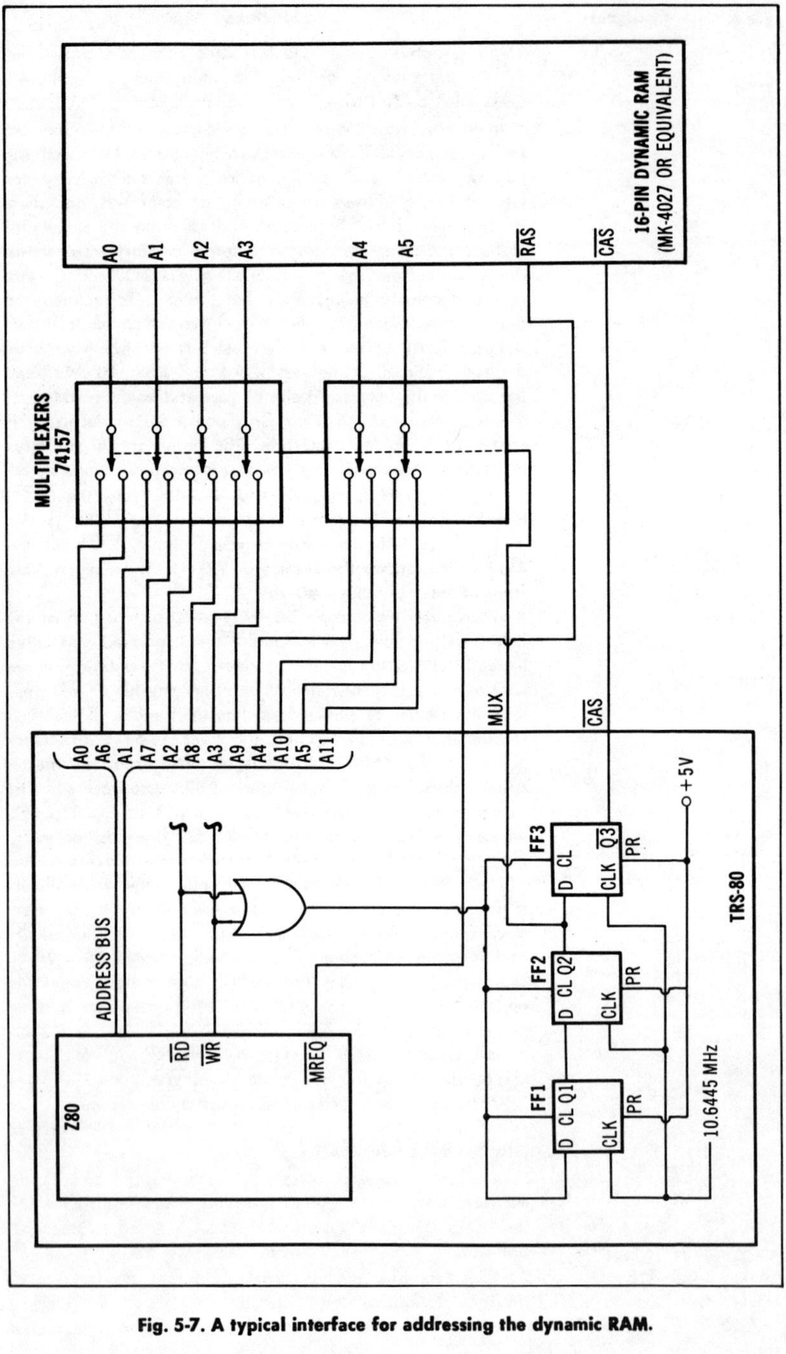

Fig. 5-7. A typical interface for addressing the dynamic RAM.

Pin No.	Signal	Definition
27, 31, 34–36, 38, 40		
21	$\overline{\text{INT}}$	Interrupt request—The interrupt request input feeds $\overline{\text{INT}}$ of the Z-80. When pulled low, the processor finishes the current instruction and examines the "I" flip-flop. If it is set, the processor pushes the program counter on the stack, resets the "I" flip-flop and begins the interrupt service routine. The processor's response to an interrupt is software selectable to one of three modes. In mode 0 the processor pulls $\overline{\text{INTAK}}$ low and expects the interrupting peripheral to place the instruction to be executed next on the data bus. In mode 1 the processor loads the program counter with hex 0066 and begins execution. In mode 2 the processor loads the program counter with the contents of two contiguous locations which are pointed to by the combination of the I register contents and an address supplied on the data bus by the interrupting device. In modes 1 and 2 the user must, obviously, have the service routine at the proper locations. If the "I" flip-flop is reset when an interrupt request occurs, the processor will ignore the request until the flip-flop is set.
14	$\overline{\text{INTAK}}$	Interrupt acknowledge—The interrupt acknowledge line is a derived output that is pulled low at the beginning of an interrupt service routine. It indicates in modes 0 and 2 that the device requesting service should place a byte on the data bus. It is derived by ORing the processor $\overline{\text{IORQ}}$ and $\overline{\text{M1}}$ output lines, which go low simultaneously to indicate an interrupt service is beginning.
33	$\overline{\text{WAIT}}$	Wait—The wait line is an active-low input that allows the processor to work with slow memory. It is sampled each time an address is put on the bus and, if pulled low by memory or a peripheral, the processor executes wait cycles until it is pulled high. Wait cycles delay the execution of an instruction, giving the memory or peripherals the extra time they need to access the desired data.
23	$\overline{\text{TEST}}$	Test—The test line is an active-low input used by peripherals to gain control of the external buses. It affects the processor bus request $\overline{\text{BUSRQ}}$ input and all on-board buffers are also floated, effectively disconnecting the TRS-80 from external devices and allowing DMA-type operations.
39	+5 V, GND	This power line is a 5-Vdc output on level I computers and a system ground on level II computers.
8, 29, 37	GND	Ground—The ground line is a common return for all system power supplies and signals.

CHAPTER 6

The TRS-80 Model III Bus

The overwhelming popularity of the original TRS-80 Model I microcomputer led Radio Shack to introduce the TRS-80 Model III, which is shown in Fig. 6-1. The Model III is Z-80 based and is software compatible with the Model I. The Model III includes an all-in-one case which houses the keyboard, processor, video monitor, ex-

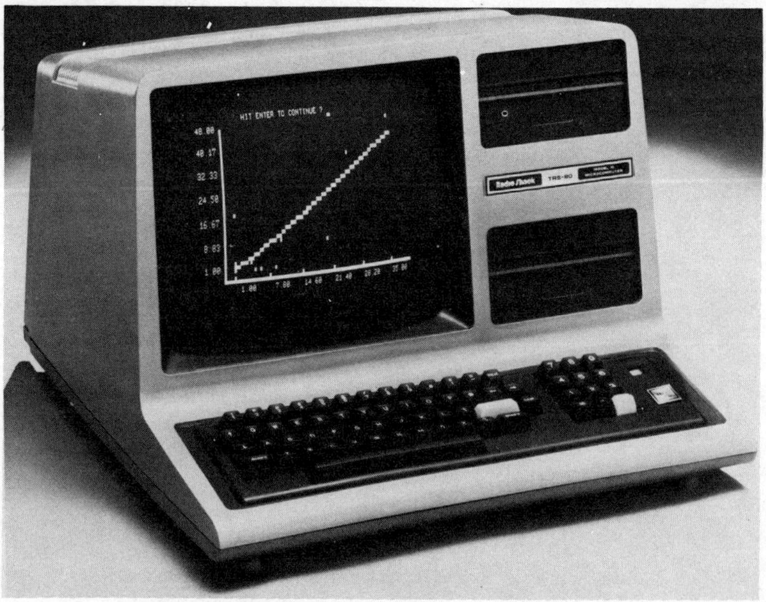

Courtesy Radio Shack Corp.

Fig. 6-1. The TRS-80 Model III microcomputer.

pansion interface circuitry, and two 5¼-inch (13.3-cm) disk drives. Memory expansion of the Model III is performed inside its case using dedicated connectors.

User-specific i/o is possible through a 50-pin edge connector located on the back of the case. Only those processor signals needed for i/o are present on the connector. The pinout of the Model III i/o bus is given in Fig. 6-2.

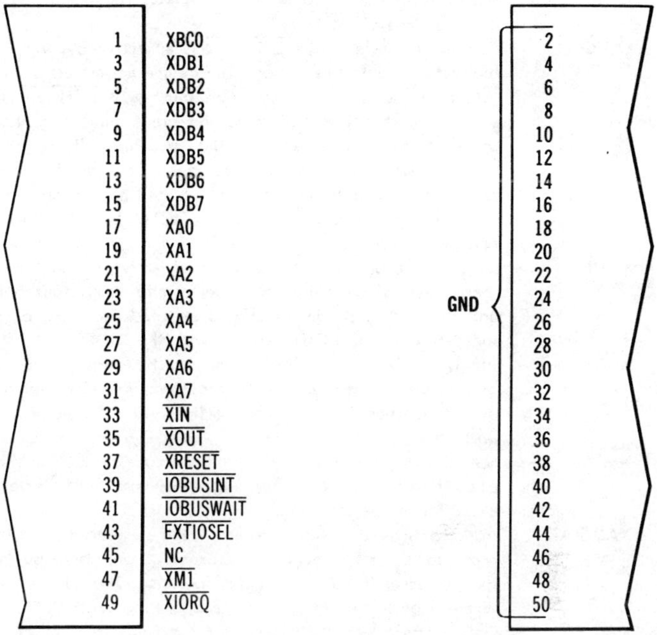

Fig. 6-2. The TRS-80 Model III i/o bus pinout.

The address, data and control lines, except $\overline{\text{IOBUSINT}}$, are fed through tri-statable buffers. The enable lines of these buffers are connected to bit 4 of output port EC, which is located on the Model III board. Bit 4 in port EC is called ENEXTIO and is configured as an output line by the operating system when the system is reset. The condition of ENEXTIO is software controllable and can be set (1) or reset (0) using the load or poke instructions. When a one (1) is stored in ENEXTIO the data, address, and control lines on the i/o bus are enabled. When a zero (0) is stored in ENEXTIO the data, address, and control lines on the i/o bus are disabled. The user must set ENEXTIO before i/o operations can be performed through the i/o bus.

TRS-80 MODEL III BUS SIGNAL DEFINITIONS

Pin No.	Signal	Definition
1, 2, 5, 7, 9, 11, 13, 15	XDB0–XDB7	Data bus—The data bus lines are buffered bidirectional lines that carry data between the processor and peripherals.
33	XIN	Input—The input line is a buffered, active-low system output that indicates the processor is performing an input operation and requests that the device addressed place data on the data bus. This line is gated with the XIORQ. The addressed device must assert EXTIOSEL.
43	EXTIOSEL	External i/o select—The EXTIOSEL is an active-low input to the bidirectional data bus buffers. The device addressed by the processor during an input operation must pull EXTIOSEL low to switch the data bus buffers to the input state and allow the processor to read the external data bus lines. If the EXTIOSEL is not pulled low during an input operation, the data bus lines remain outputs and valid data is not read by the processor. An input device can easily generate EXTIOSEL by NANDing the IN and port address lines.
35	XOUT	Output—The output line is a buffered, active-low system output that indicates the processor is performing an output operation and the addressed device should be ready to latch data from the data bus. The XOUT is gated with the XIORQ.
49	XIORQ	Input/output request—The XIORQ is a buffered, active-low processor output that indicates the processor is performing an input or output operation and address lines A0–A7 hold a valid i/o port address.
47	XM1	M1—The M1 is a buffered, active-low processor output that indicates the processor is performing an op-code fetch cycle during execution of an instruction.
17, 19, 21, 23, 25, 27, 29, 31	XA0–XA7	Address lines 0–7—The address lines are buffered, active-low processor outputs used to indicate which port is to be used by the processor in the current i/o operation. The address lines indicate a port address an and i/o operation when IORQ is low. Only ports 80–FF are reserved for system use; all others are available to the user.
37	XRESET	Reset—The reset is a buffered, active-low system output that is pulled low during a processor reset.
39	IOBUSINT	I/o bus interrupt request—The IOBUSINT is a buffered, active-low input to the processor's INT input. The Model III supports mode 1 interrupts. The user must store the starting address vector of the interrupt service routine at 403E and 403F before the interrupt request occurs. The starting address is usually stored during system initialization. The IOBUSINT is enabled by bit 3, called ENIOBUSINT, an output line in output port EO. When bit 3 of output port EO is high (1) the IOBUSINT is enabled and the processor may recognize an interrupt request from a device connected to the IOBUS. When bit 3 of OUTPUT EO is low (0) the IOBUSINT is disabled and an interrupt request will not be recognized by the processor. The status of IOBUSINT can be read any time by examining bit 3 in input port EO. If bit 3 in input port EO is low (0) IOBUSINT has been pulled low by an external device requesting service.

Pin No.	Signal	Definition
41	$\overline{\text{IOBUSWAIT}}$	I/o bus wait—The i/o bus wait is a buffered, active-low processor input to the Z-80's wait line. When the wait line is pulled low during an input or output operation, by a slow peripheral, the processor inserts wait states, giving the peripheral extra time to place data on or latch data from the data bus. Care must be exercised in using the $\overline{\text{IOBUSWAIT}}$ because the Model III uses dynamic memory devices which depend on the processor for refresh signals. Holding the $\overline{\text{IOBUSWAIT}}$ low for longer than 500 μs may endanger the data stored in RAM.

CHAPTER 7

The TRS-80 Model II Bus

Radio Shack expanded its line of computers with the introduction of its business system, the TRS-80 Model II, shown in Fig. 7-1. The Model II is Z-80 based, includes one 8-inch (20.3-cm) ½-megabyte disk drive, 32K or 64K of RAM, two serial ports, and one parallel printer port.

Courtesy Radio Shack Corp.

Fig. 7-1. The TRS-80 Model II microcomputer.

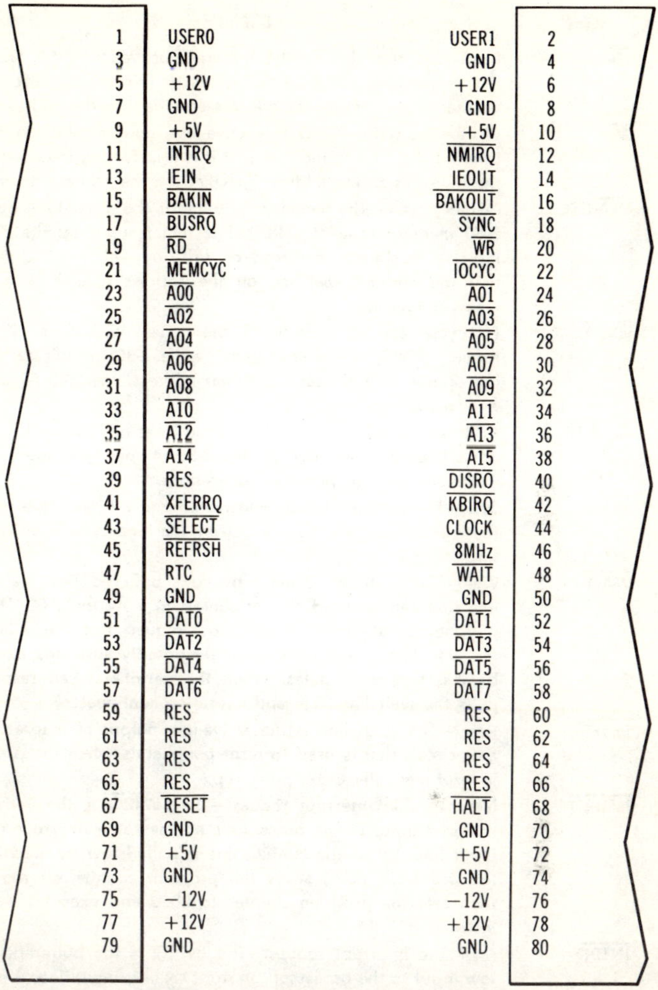

Fig. 7-2. The TRS-80 Model II bus pinout.

The Model II features an internal motherboard with four empty slots. The motherboard accepts 80-pin double-edge printed-circuit cards. The pinout of the Model II bus is shown in Fig. 7-2.

TRS-80 MODEL II SIGNAL DEFINITIONS

Pin No.	Signal	Definition
51–58	$\overline{D0}$–$\overline{D7}$	Data bus—The data bus is the inverted, buffered, bidirectional data lines from the Z-80 which carry data and instructions between the processor and peripherals.

Pin No.	Signal	Definition
19	$\overline{\text{RD}}$	Read—The read line is the buffered, active-low Z-80 read line that is used by the Z-80 to indicate it is receiving information on the data bus from the address specified on the address lines.
20	$\overline{\text{WR}}$	Write—The write line is the active-low, buffered Z-80 write line used by the Z-80 to indicate that it is transferring information on the data bus to the address specified on the address lines.
21	$\overline{\text{MEMCYC}}$	Memory cycle—The memory cycle line is the active-low, buffered Z-80 memory request (MREQ) line that is used by the Z-80 to indicate that a memory read or write operation will occur and that the address specified on the address lines is a memory mapped location.
22	$\overline{\text{IOCYC}}$	I/o cycle—The i/o cycle line is the active-low, buffered Z-80 i/o request (IORQ) line that is used by the Z-80 to indicate that an i/o operation will occur and that address lines A0–A7 specify an i/o port address.
23–38	$\overline{\text{A0}}$–$\overline{\text{A15}}$	Address bus—The address bus is the inverted, buffered Z-80 address lines that are used by the Z-80 to specify memory or i/o addresses used in processor operations.
18	$\overline{\text{SYNC}}$	Sync—The sync is the active-low, buffered Z-80 M1 line used by the Z-80 to indicate that the current machine cycle is an op-code fetch.
48	$\overline{\text{WAIT}}$	Wait—The wait line is the active-low, buffered Z-80 wait input line that can be used by peripherals to slow the Z-80. When a peripheral that cannot respond to a request from the Z-80 pulls the wait line low the processor temporarily suspends execution by inserting wait states. When the peripheral can respond it pulls the wait line high and execution continues.
67	$\overline{\text{RESET}}$	Reset—The reset line is the active-low output of a reset generator circuit that is used to reset peripherals when the processor is reset manually or on power-up.
12	$\overline{\text{NMIRQ}}$	Nonmaskable interrupt request—The $\overline{\text{NMIRQ}}$ is the active-low, buffered input to the processor's nonmaskable interrupt request input line. When the $\overline{\text{NMIRQ}}$ line is pulled low by a peripheral the processor halts, saves the program counter on the stack, then sets the program counter to 0066 and executes a restart instruction.
11	$\overline{\text{INTRQ}}$	Maskable interrupt request—The $\overline{\text{INTRQ}}$ is the buffered, active-low input to the processor interrupt request input line. When the $\overline{\text{INTRQ}}$ line is pulled low by a peripheral the processor responds in one of three modes. (See TRS-80 Model I INT definition for explanation of modes.)
13, 14	IEIN, IEOUT	Interrupt enable input and interrupt enable output—The IEIN and IEOUT lines are used to form an interrupt priority daisy chain. The peripheral board that is physically closest to the processor, board 0, has its IEIN line tied high. Its IEOUT line is connected to the IEIN line on the next board, board 1. That board's IEOUT line is connected to the IEIN line of board 2, and so on. The i/o boards must contain circuitry to evaluate the status of IEIN and set IEOUT. If IEIN is high, the on board interrupt can be enabled, or if the board has not requested interrupt service then it sets IEOUT high, enabling the next lower priority interrupt. If IEIN is low, a higher priority board is also requesting

Pin No.	Signal	Definition
		interrupt service so the on-board interrupt cannot be enabled and its IEOUT line must be held low.
17	$\overline{\text{BUSRQ}}$	Bus request—The bus request is the active-low, buffered Z-80 bus request input used by peripherals to gain control of the buses. When pulled low by a peripheral, the processor suspends execution and then floats the buses and pulls the bus acknowledge line low. An external device may then use the buses for DMA or other operations.
41	$\overline{\text{XFERRQ}}$	DMA transfer request input—The $\overline{\text{XFERRQ}}$ is the ready input line on the Z-80 DMA device. When an external device pulls the XFERRQ line low the DMA device initiates a DMA sequence of operations.
15, 16	$\overline{\text{BAKIN}}$ $\overline{\text{BAKOUT}}$	Bus acknowledge input and bus acknowledge output—The $\overline{\text{BAKIN}}$ and $\overline{\text{BAKOUT}}$ are active-low lines that form a DMA priority daisy chain structure like that formed by the IEIN and IEOUT lines. The board physically closest to the processor, 0, has its $\overline{\text{BAKIN}}$ fed by the processor's bus acknowledge line. The $\overline{\text{BAKOUT}}$ line on board 0 feeds the $\overline{\text{BAKIN}}$ line on board 1, etc. Each board must be able to interpret the $\overline{\text{BAKIN}}$ and set the condition of $\overline{\text{BAKOUT}}$.
42	$\overline{\text{KBIRQ}}$	Keyboard interrupt request—The $\overline{\text{KBIRQ}}$ is an active-low input to the Z-80 CTC device.
43	$\overline{\text{SELECT}}$	Select—The $\overline{\text{SELECT}}$ is an active-low output used by the processor to enable and disable memory boards in future applications.
40	$\overline{\text{DISRO}}$	Disable RAM output—The $\overline{\text{DISRO}}$ is an active-low input to memory management circuitry used by the system to allow memory overlays in future applications.
68	$\overline{\text{HALT}}$	Halt—The halt is a buffered, active-low processor output that is pulled low by the processor to indicate that it has executed a halt instruction and is waiting for an interrupt.
45	$\overline{\text{REFRESH}}$	Refresh—The refresh line is a buffered, active-low processor output used to refresh dynamic memories.
44	CLOCK	Clock—The clock is the buffered 4-MHz system clock.
46	8 MHz	8 MHz—The 8 MHz is a buffered 8-MHz clock output.
47	RTC	Real-time clock heartbeat—The RTC is a buffered 60-hertz clock signal that can be used in timing applications.
1, 2	USR0, USR1	User definable lines—USR0 and USR1 are not specified and may be defined by the user for special applications.
39, 59–66	RES	Reserved—These reserved lines are reserved for future use by Radio Shack.
9, 10 71, 72	+5V	+5 volts—The +5-V lines are +5-volt dc regulated voltage outputs.
5, 6, 77, 78	+12V	+12 volts—The +12-V lines are +12-volt dc regulated outputs.
75, 76	−12V	−12 volts—The −12-V lines are −12-volt dc regulated outputs.
3, 4, 7, 8, 49, 50, 69, 70, 73, 74, 79, 80	GND	Ground—The GND lines are system ground lines.

CHAPTER 8

The TRS-80 Color Computer

Radio Shack's TRS-80 Color Computer is the first of the new generation of high-performance, low-cost, personal microcomputers. The TRS-80 Color Computer, shown in Fig. 8-1, includes an alphanumeric keyboard, color video generator, BASIC in ROM, instant programmability through a ROM cartridge slot, a cassette interface, and an RS-232 serial interface.

The cartridge slot accepts ROM-based plug-in cartridges that contain prepackaged user programs. The bus lines brought to the cartridge connector, a double-side edge card connector, also allow memory and i/o expansion. The pinout of the cartridge connector is shown in Fig. 8-2. Unlike the Z-80-based TRS-80 Models I, II, and III, the TRS-80 Color Computer is based on Motorola's 6809 microprocessor, and signals on the bus are 6809 derived.

TRS-80 COLOR COMPUTER SIGNAL DEFINITIONS

Pin No.	Signal	Definition
10–17	D0–D7	Data bus—The data bus is the buffered, bidirectional 6809 data lines. The data bus carries instructions and data to and from the 6809.
18	$R/\overline{W}$	Read/write—The read/write line is the buffered 6809 $R/\overline{W}$ line that is used by the 6809 to control direction of data flow on the data bus. When the $R/\overline{W}$ is high the 6809 is reading data from a peripheral or memory; when the $R/\overline{W}$ is low the 6809 is writing data to a peripheral or memory.
6	E	Enable—The enable is the 0.89-MHz buffered 6809 system $\phi 2$ clock line. The enable is used by memory and peripherals to latch data from or place data on the data bus.
7	$\overline{Q}$	Quadrature—The quadrature is the buffered 6809 $\overline{Q}$ clock line. The $\overline{Q}$ signal leads the enable signal by 90° and indicates that address and data placed on the bus by the processor are stable.

Pin No.	Signal	Definition
19–31, 37–39	A0–A15	Address bus—The address bus is the buffered 6809 A0–A15 lines. The address lines are processor outputs used by the 6809 to select the memory or i/o location needed in the current operation.
8	$\overline{\text{CART}}$	Cartridge—The cartridge line is normally used by the system to see if a cartridge has been inserted. $\overline{\text{CART}}$ is an active-low input to a peripheral interface adapter interrupt input, which feeds the 6809's fast interrupt request ($\overline{\text{FIRQ}}$) input line. When a ROM cartridge is inserted into the cartridge connector, circuitry in the cartridge shorts the processor's Q output to the $\overline{\text{CART}}$ line. During the next 6809 clock cycle the falling edge of the Q signal pulls the $\overline{\text{FIRQ}}$ low, through a PIA interrupt line. The processor responds to the $\overline{\text{FIRQ}}$ by pushing only the program counter and condition code register onto the stack, masking further $\overline{\text{FIRQ}}$ requests, and jumping to location C000 to begin the interrupt service routine. When a program cartridge has been inserted, the service routine is the program located in the cartridge ROMs. The $\overline{\text{CART}}$ can be used as general $\overline{\text{FIRQ}}$ input that has a service routine located at C000.

Courtesy Radio Shack Corp.

Fig. 8-1. The TRS-80 Color Computer.

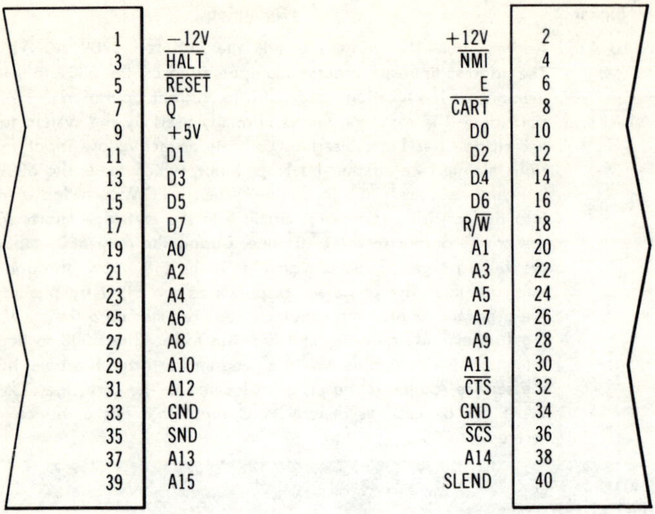

Fig. 8-2. Pinout designations for the TRS-80 Color Computer bus.

Pin No.	Signal	Definition
3	$\overline{\text{CTS}}$	Cartridge select—The cartridge select is a buffered, active-low system output that is used to enable memory or i/o devices located between C000 and FF00. The $\overline{\text{CTS}}$ is generated after a cartridge has been inserted and the $\overline{\text{CART}}$ sent to the processor which then asserts $\overline{\text{CTS}}$ and enables the cartridge ROMs. It is also asserted when any device in the C000–FF00 block is addressed during program execution.
36	$\overline{\text{SCS}}$	Spare chip select—The spare chip select is a buffered, active-low system output that is used to enable memory or i/o devices located between FF00 and FF40.
5	$\overline{\text{RESET}}$	Reset—The reset is the buffered, active-low output from the system reset generation circuitry. The $\overline{\text{RESET}}$ is used to initialize the system upon power-up and at the operator's request.
4	$\overline{\text{NMI}}$	Nonmaskable interrupt—The nonmaskable interrupt is an active-low input to the 6809's $\overline{\text{NMI}}$ line. When pulled low by a peripheral, the $\overline{\text{NMI}}$ forces the processor to cease execution at the end of the current program, push the register contents onto the stack, mask $\overline{\text{FIRQ}}$ and $\overline{\text{IRQ}}$ requests, and begin execution of the service routine.
3	$\overline{\text{HALT}}$	Halt—The halt line is an active-low input to the 6809's $\overline{\text{HALT}}$ line. When a peripheral pulls the $\overline{\text{HALT}}$ low the processor ceases execution at the end of the current operation and floats the data, address, and control bus lines. The peripheral may then use the buses for DMA or multiprocessor operations.
40	$\overline{\text{SLEND}}$	Select enable—The select enable is an active-low system input used by external devices, e.g., cartridges, to disable internal Color Computer device selection circuitry. When an external device pulls the select enable low it can access any device on the bus located inside the Color Computer.
35	SND	Sound—The sound line is an input to the mixer in the video generator and allows sound input from cartridges or other devices operating in the cartridge slot.

Pin No.	Signal	Definition
9	+5V	+5 volts—The +5-V line supplies up to 300 mA of regulated +5 volts dc.
2	+12V	+12 volts—The +12-V line supplies up to 300 mA of regulated +12 volts dc.
1	−12V	−12 volts—The −12-V line supplies up to 100 mA of regulated −12 volts dc.
33, 34	GND	Ground—The ground line is the system signal and power ground.

CHAPTER 9

The SS-50/C Bus

The SS-50 bus was introduced in 1975 by Southwest Technical Products Corporation (SWTPC) for their SWTPC 6800 microcomputer system. Figs. 9-1 and 9-2 show the system. Since then, use of the SS-50 has grown so that, today, at least a dozen manufacturers produce components for it that range from disk controllers to digital video boards. All manufacturers of the SS-50 conform to the original bus definitions laid down by Southwest Technical Products Corporation. The compatibility problems that plague the S-100 bus do not exist for the SS-50 bus.

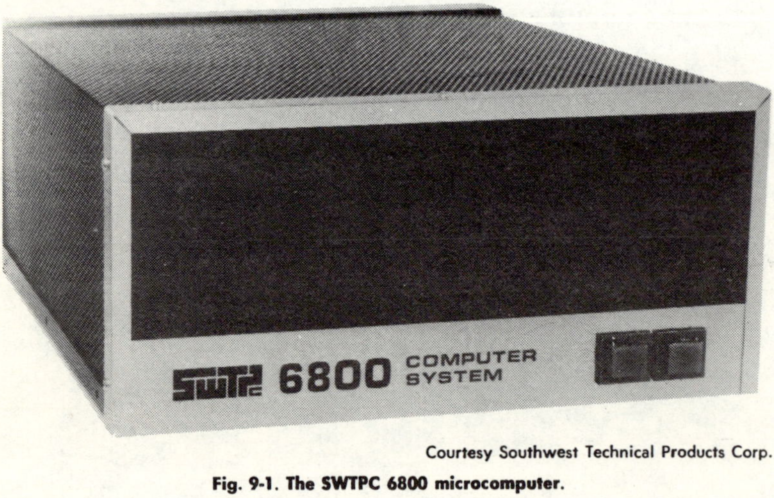

Courtesy Southwest Technical Products Corp.

Fig. 9-1. The SWTPC 6800 microcomputer.

Since it was based on the SWTPC 6800 microcomputer system, the SS-50 bus easily supports the 6800 and 6502 processors. Unlike the 8080, these processors do not multiplex control signals on the data bus. All control signals are derived on the processor. Parallel and serial i/o ports and memory are all treated alike by the CPU. Each is addressable and must respond to a simple R/W line and the address bus.

Courtesy Southwest Technical Products Corp.

Fig. 9-2. A complete SS-50 system.

Most of the signals on the SS-50 bus come directly from the 6800 processor. The SS-50 data bus is bidirectional; it is not split into data-in and data-out lines.

The SWTPC motherboard, shown in Fig. 9-3, accepts two different size boards—an SS-50 board (shown in Fig. 9-4) and a peripheral board called the SS-30 (shown in Fig. 9-5). The SS-30 board uses a subset of SS-50 signals. Each board will be discussed in turn.

SS-50 SIGNAL DEFINITIONS

The SS-50 boards are 9 by 5½ inches (22.8 by 13.9 cm) in size. They plug into the motherboard through 50-pin Molex connectors as shown in Figs. 9-3 and 9-4. The pins are numbered from right to

Courtesy Southwest Technical Products Corp.

Fig. 9-3. The SWTPC 6809 motherboard.

Courtesy The Micro Works

Fig. 9-4. An SS-50 PROM board.

Courtesy The Micro Works

Fig. 9-5. An SS-30 EPROM programmer board.

left. Positive indexing is provided by plugging pin 33 on all boards. The SS-50 pinout designations are listed in Fig. 9-6.

Pin No.	Signal	Definition
1–8	D0–D7	Data bus—The SS-50 data bus is the complement of the 6800 data bus lines D0–D7. These lines are bidirectional and carry information between the processor and all peripherals.
41	R/$\overline{\text{W}}$	Read/write—The read/write line of the processor, connecting the processor and peripherals, determines the direction of data flow on data bus lines D0–D7. When it is high, data flows into the processor in a read operation. When it is low, data flows out of the processor in a write operation.
9–24	A0–A15	Address lines—The address bus lines of the processor connect the processor and peripherals, thereby providing the desired address.
40	$\overline{\text{VMA}}$	Valid memory address—This valid memory-address line is the complemented 6800 VMA line. It goes low when a valid address has been placed on the address bus. It is often used to enable peripherals.
34	$\overline{\text{M RESET}}$	Manual reset—The manual reset line is the input to a one-shot multivibrator. When pulled low by pressing the reset button, the one-shot develops a pulse on the reset line which resets the system.
42	$\overline{\text{RESET}}$	Reset—The reset line is the output of the one-shot triggered either by the $\overline{\text{M RESET}}$ or automatically on system power up. It is connected to the reset input of the processor, and to other resettable peripherals. The processor responds to this line by loading the program counter with the contents of the reset vector and beginning execution.
35	$\overline{\text{NMI}}$	Nonmaskable interrupt—The nonmaskable interrupt is an active-low line connected to the processor's $\overline{\text{NMI}}$ input. The processor responds to this line by halting execution of the current program,

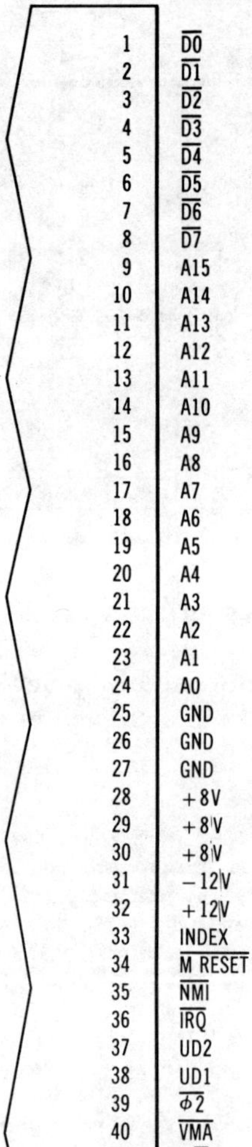

Fig. 9-6. The SS-50 pinout designations.

Pin No.	Signal	Definition
		storing the vital registers on the stack, loading the program counter with the contents of the $\overline{NMI}$ vector and beginning execution of the $\overline{NMI}$ service routine. The processor must respond to an input on this line.
36	$\overline{IRQ}$	Interrupt request—This is an active-low line connected to the processor's $\overline{IRQ}$ input. The processor can, if the interrupt status flag is set, ignore an input on this line. If the "I" flag is reset, it can respond to an input by halting execution of the current program, storing the vital registers on the stack, loading the program counter with the contents of the $\overline{IRQ}$ vector and beginning execution of the $\overline{IRQ}$ service routine.
45	$\overline{HALT}$	Halt—The halt is an active-low line connected to the halt input of the processor. The processor responds to this line by halting execution of the current program and floating the address and data buses and the R/W line. An external device may then access memory for DMA or other peripherals.
43	BA	Bus available—This is an output connected to the BA line of the processor. Although normally low, it goes high in response to a halt input signalling an external device that the buses are available for DMA.
44	$\phi1$	$\phi1$—The $\phi1$ line is the $\phi1$ output from the two-phase nonoverlapping clock of the processor.
39	$\overline{\phi2}$	$\phi2$—The $\overline{\phi2}$ output is an inverted $\phi2$ output from the processor's two-phase nonoverlapping clock and is used to indicate the presence of valid data on the data bus.
37, 38	UD1, UD2	User defined—User defined lines 1 and 2 are common to all SS-50 boards but are not assigned functions. They allow the user to customize by assigning them special functions in his system.
46–50	110b, 150b, 300b, 600b, 1200b	Serial clock—The serial clock lines are the outputs of a crystal-controlled baud-rate generator and can be used by a serial interface peripheral.
25–27	GND	Ground—The ground line is the common return line for all of the system power supplies.
28–30	+8V	+8 volts—The +8-V line is an unregulated power-bus input that can be used by on-board regulators to develop +5 volts.
31	−12V	−12 volts—The −12-V line is a regulated negative 12-volt input.
32	+12V	+12 volts—The +12-V line is a regulated positive 12-volt input.
33	INDEX	Index—The index line is physically plugged to prevent incorrect insertion of a system board.

THE SS-30 PERIPHERAL BOARD

The SS-30 is a subset of SS-50 signals designed to facilitate use of both serial and parallel interface devices. The most obvious difference between the two is the absence of the address bus on the SS-30. It is replaced by board select lines which are derived from the SS-50 address bus. The SWTPC 6800 motherboard accommodates eight peripheral boards on the SS-30. Each board has been assigned a block of four contiguous addresses in memory as shown in Fig. 9-7. Address decoding circuitry on the motherboard generates

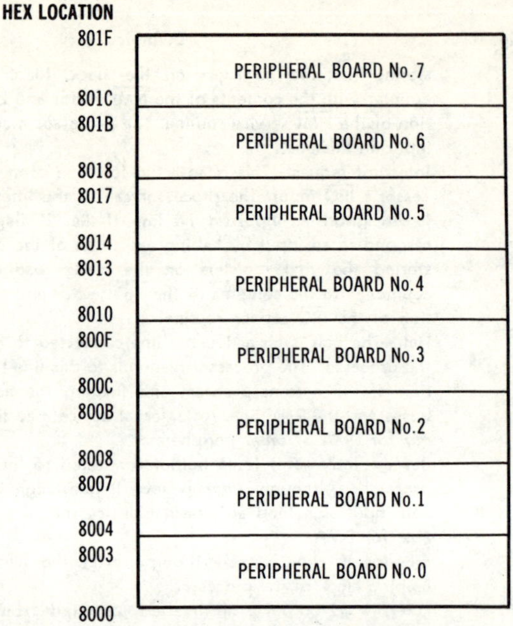

Courtesy Southwest Technical Products Corp.

Fig. 9-7. Memory locations of the SS-30 interface boards.

a board select signal (using a 1-of-8 decoder) when the base address assigned a board is put on the address bus. Since the SWTPC motherboard can accommodate eight peripheral boards, there are eight board select lines. The board select signal is fed only to the proper board, which must then respond to the register select lines, RS0 and RS1. The RS0 and RS1 lines are connected to all peripheral boards and represent address bus lines A0 and A1. Using the RS0 and RS1 lines, a board can determine which of its four locations is being addressed. One of the peripheral boards is usually dedicated to serial communications with an external terminal. SS-30 boards are 5 by 4 inches (12.7 by 10.1 cm) and plug into the motherboard through 30-pin connectors as shown in Figs. 9-3 and 9-5. Positive indexing is provided by plugging pin 7.

SS-30 SIGNAL DEFINITIONS

The SS-30 pinout designations are listed in Fig. 9-8.

Pin No.	Signal	Definition
1, 2	UD3, UD4	User defined 3 and 4—These lines are common to all the SS-30 boards but not assigned functions. They are available to the user to allow special individualized functions. They are not connected to UD1 and UD2 on the SS-50.

Pin No.	Signal	Definition
3, 4	−12V, +12V	±12 volts—The ±12-V supply from the SS-50.
5, 6	GND	Ground—The ground line from the SS-50.
7	INDEX	Index—The index line is physically plugged to prevent incorrect insertion of a peripheral board.
8, 9	$\overline{NMI}$, $\overline{IRQ}$	Nonmaskable interrupt and interrupt request—These are the nonmaskable interrupt and interrupt request lines from the SS-50.
10, 11	RS0, RS1	Register select—Register select lines 1 and 2 are connected to SS-50 address bus lines A0 and A1, respectively. They are used by a peripheral board (when signalled by its board select line) to determine which of its four addresses is being accessed.
12–19	D0–D7	Data lines—Data lines D0–D7 are inverted $\overline{D0}$–$\overline{D7}$ lines from the SS-50.
20	$\phi2$	$\phi2$—The $\phi2$ line is the inverted $\overline{\phi2}$ line from the SS-50.
21	R/$\overline{W}$	Read/write—This is the read/write line from the SS-50.
22, 23	+8V	+8 volts—This is the +8-V unregulated supply from the SS-50.
24–28	1200b, 600b, 300b, 150b, 110b	Serial clock—These are the serial clock lines from the SS-50.

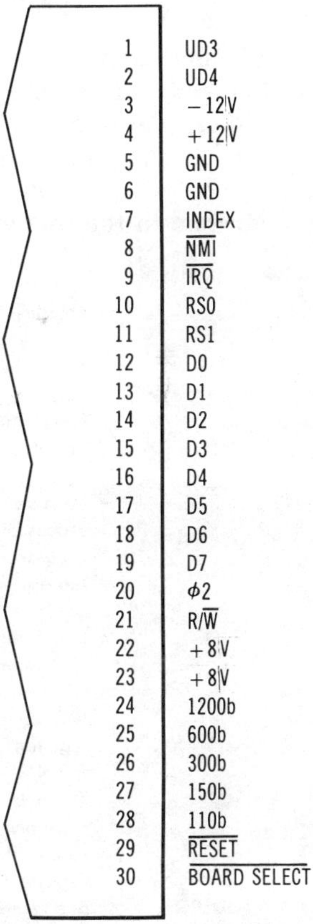

Fig. 9-8. The SS-30 pinout designations.

Pin No.	Signal	Definition
29	$\overline{\text{RESET}}$	Reset—This is the reset line from the SS-50.
30	$\overline{\text{BOARD SELECT}}$	Board select—There are eight board select lines, one to each peripheral board. The select lines are active low. Thus the board line selected is low while the others are held high. Only one board can be selected at a time. The board select lines are decoded from the SS-50 address bus. The address of each board follows:

Board and Line Number	Lowest Hex Address
SEL 0	8000
SEL 1	8004
SEL 2	8008
SEL 3	800C
SEL 4	8010
SEL 5	8014
SEL 6	8018
SEL 7	801C

SS-50C SIGNAL DEFINITIONS

Implementation of Motorola's 6809 microprocessor on the SS-50 required redefinition of several lines. Most of the changes resulted from differences in control lines between the 6800 and the 6809. The redefined bus is known as the SS-50C. The SS-50C pinout designations are listed in Fig. 9-9. Only those lines that have been redefined are specified in the following information.

Pin No.	Signal	Definition
39	$\overline{\text{E}}$	Enable—The enable line is the buffered, active-low processor output that is the system clock, formerly $\phi 2$. Like the $\phi 2$ line on the 6800, the enable line is used by memory and peripherals to latch data from or place data on the data lines.
38	$\overline{\text{Q}}$	Quadrature—The $\overline{\text{Q}}$ line is the buffered, active-low processor output that leads the enable signal by 90° and signifies that addresses and data, placed on lines by the processor, are stable.
34	$\overline{\text{MRDY}}$	Memory ready—The memory ready is the buffered, active-low processor input used by peripherals and memory to slow the processor. Pulling this line low causes the processor to stretch the enable output to a maximum of 10 μs, allowing slow memory extra time to place data on or latch data from the data lines.
37	$\overline{\text{FIRQ}}$	Fast interrupt request—The $\overline{\text{FIRQ}}$ is the buffered, active-low fast $\overline{\text{IRQ}}$ input to the processor. The $\overline{\text{FIRQ}}$ is maskable; the flag in the CCR holds the mask condition. The processor responds to a fast $\overline{\text{IRQ}}$ by pushing only the program counter and condition code register (CCR) onto the stack, clearing the E bit in the CCR, fetching the $\overline{\text{FIRQ}}$ vector and beginning the service routine. The E bit is used by an RTI instruction to determine how many registers are to be pulled from the stack. Processor response to an $\overline{\text{FIRQ}}$ is fast because only the PC and CCR, not the entire register set, are saved.
46	$\overline{\text{BUSRQ}}$	Bus request—The $\overline{\text{BUSRQ}}$ is the buffered, active-low BREQ input.

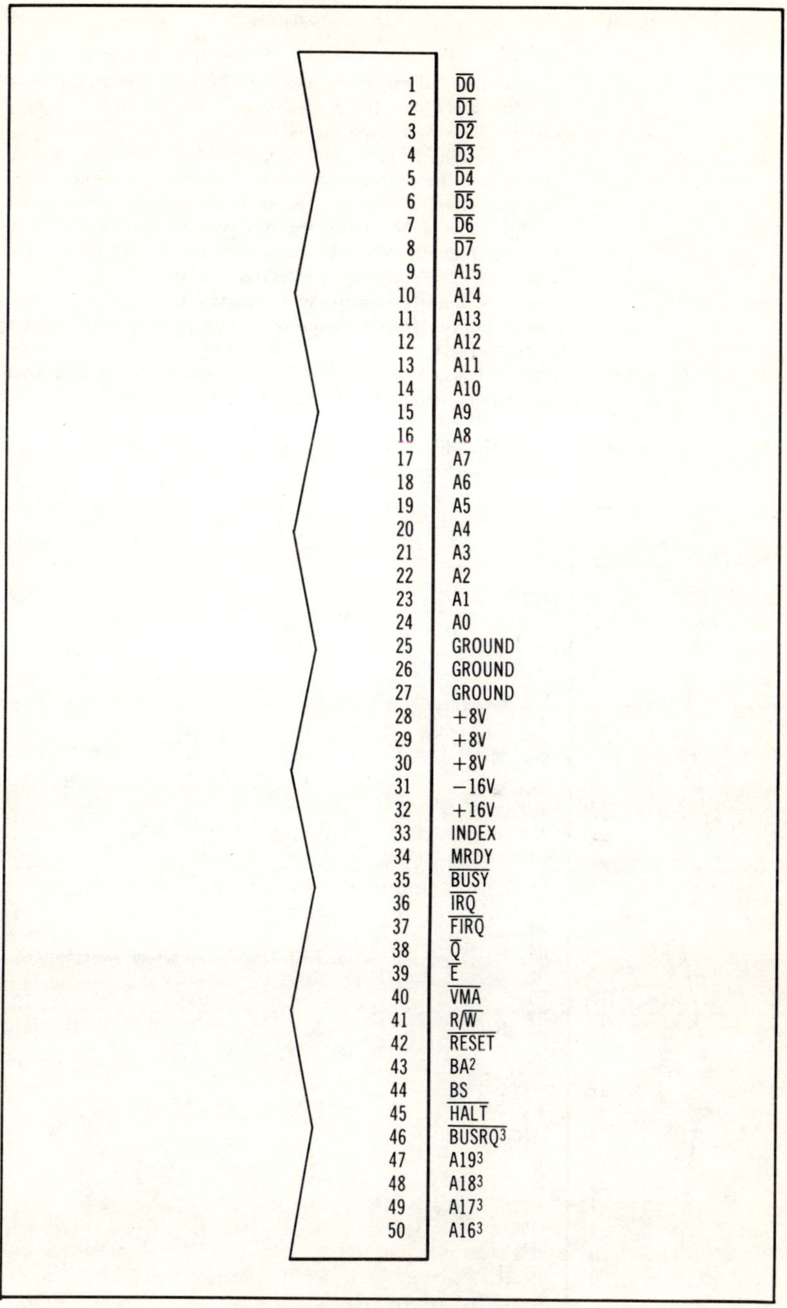

Fig. 9-9. The SS-50C pinout designations.

Pin No.	Signal	Definition
		When pulled low by a peripheral the processor responds by floating the address, R/W, and data lines for memory refresh or DMA operations. The processor pulls the BA high to signify that the lines have been floated.
44	BS	Bus status—The BS is the buffered, active-high bus status processor output. The processor pulls the BS high to indicate that a halt, bus grant, or interrupt service is in progress. When the BS is high and the BA is low the processor is acknowledging an interrupt request. When the BS is high and the BA is high the processor is acknowledging a halt or bus grant.
35	$\overline{BUSY}$	Busy—The busy is an active-low processor output that is pulled low during read/modify/write operations to deny other masters access to the bus.
50–47	A16–A19 or S0–S3	Address lines 16–19—The S0–S2 lines are extended address lines that may be implemented on the bus.
31, 32	−16V, +16V	−16 volts, +16 volts—The −16-V and +16-V lines are unregulated power supply lines.

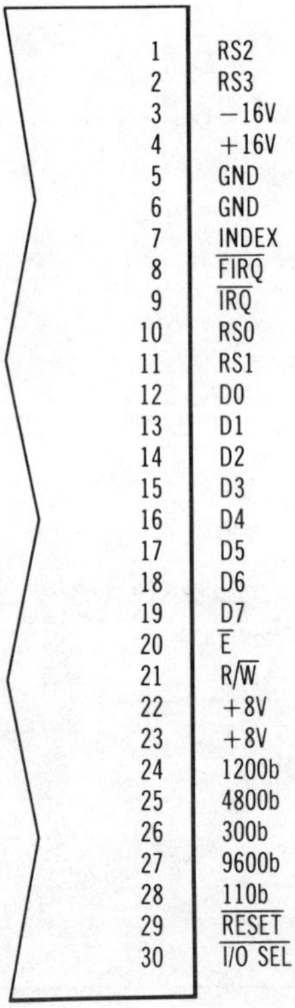

Fig. 9-10. The SS-30C pinout designations.

SS-30C SIGNAL DEFINITIONS

Some of the revisions of the SS-50 were applied on the SS-30. The SS-30C pinout is given in Fig. 9-10. Only those lines that have been refined are specified in the following information.

Pin No.	Signal	Definition
20	$\overline{E}$	Enable—The enable line is a buffered and inverted SS-50C enable line.
1, 2	RS2, RS3	Register select 2, 3—The register select 2 and 3 lines are the buffered SS-50C A2 and A3 lines.
8	$\overline{FIRQ}$	Fast interrupt request—The SS-50C $\overline{FIRQ}$ line.
25	4800b	4800 baud—The 6800 baud is a 4800-baud clock for serial interfaces in the system.
27	9600b	9600 baud—The 9600 baud is a 9600-baud clock for serial interfaces in the system.
3, 4	−16V, +16V	−16 volts, +16 volts—The −16-V and +16-V SS-50C power lines.

CHAPTER 10

The KIM/AIM/SYM Bus

The KIM-1 microcomputer shown in Fig. 10-1 was introduced in 1975 by MOS Technology, Inc. The KIM-1 is a single-board system that includes a 6502 processor, a monitor in ROM, 1 kilobyte of user accessible RAM, a hexadecimal keypad and display, 15 parallel i/o lines, and cassette and TTY interfaces. Although it has been very popular, relatively few plug-compatible expansion boards have been produced for the KIM-1.

Because it can stand alone as a complete system, the KIM-1 has no built-in motherboard into which expansion boards can be plugged. Expansion is performed via edge-card connectors to which the buses of the system are brought. MOS Technology, Inc., has introduced an expansion motherboard, the KIM-4, that plugs onto the KIM-1 edge connectors and provides connections for six expansion boards. The arrangement of lines on the motherboard is the official KIM bus. Since expansion is possible through either the KIM-1 expansion connector or the KIM-4 motherboard, both will be defined.

The AIM and SYM systems include expansion connectors that are generally compatible with the KIM-1 expansion connector. The differences are slight and occur on lines E16–E20 and E–X. These lines are system specific and should not affect operation of many boards.

Expansion of the KIM-1 using either connector is straightforward as the 6502 is much like the 6800 in its simplicity. Memory and i/o must respond alike to the address and data buses and to the read/write line. Care must be taken in assigning memory or peripheral addresses since the KIM-1 uses two memory areas as shown in Fig. 10-2—1 kilobyte of RAM rests from 0000 to 03FF and the operating system occupies 1700 to IFFF. Select signals for 4 kilobytes of expansion (0400 to 13FF) are decoded on the KIM-1 and available at the application connector. The interrupt vectors of the system are hardware located from 17FA to 17FF, the highest addresses in the KIM-1. Expansion beyond the initial 8K must include hardware to define the locations of those vectors.

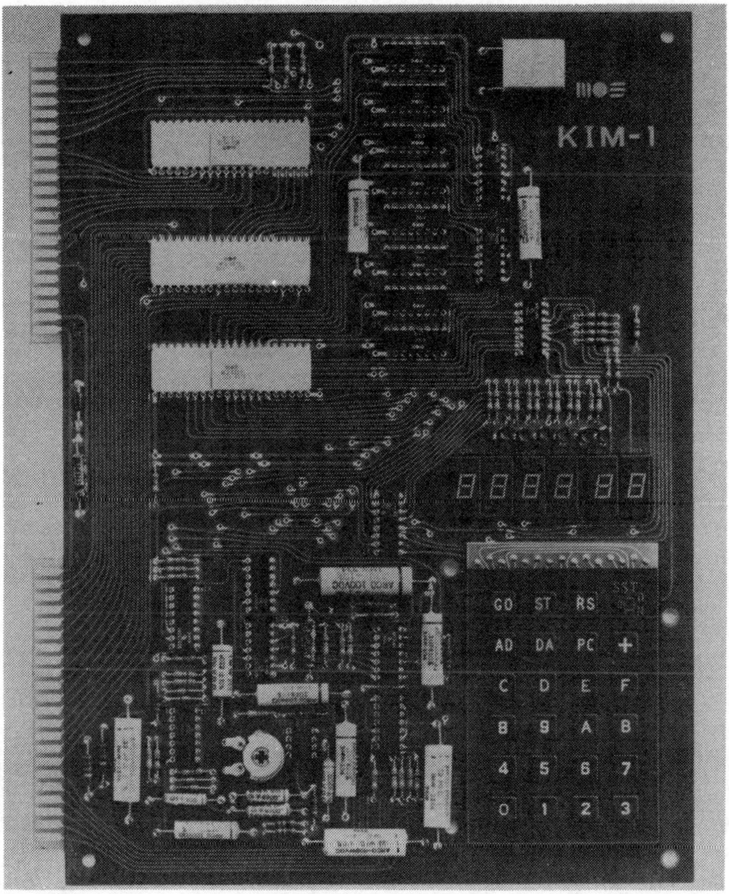

Courtesy MOS Technology, Inc.

Fig. 10-1. KIM-1 microcomputer.

The data, address and control buses are brought to the KIM-1 expansion connector but few of these lines are buffered. All KIM-1 and KIM-4 boards use 44-pin (22 each side) edge-card connectors (Vector R644). Positive indexing was not provided until August 1978, when an indexing slot was designated between pins 18 and 19.

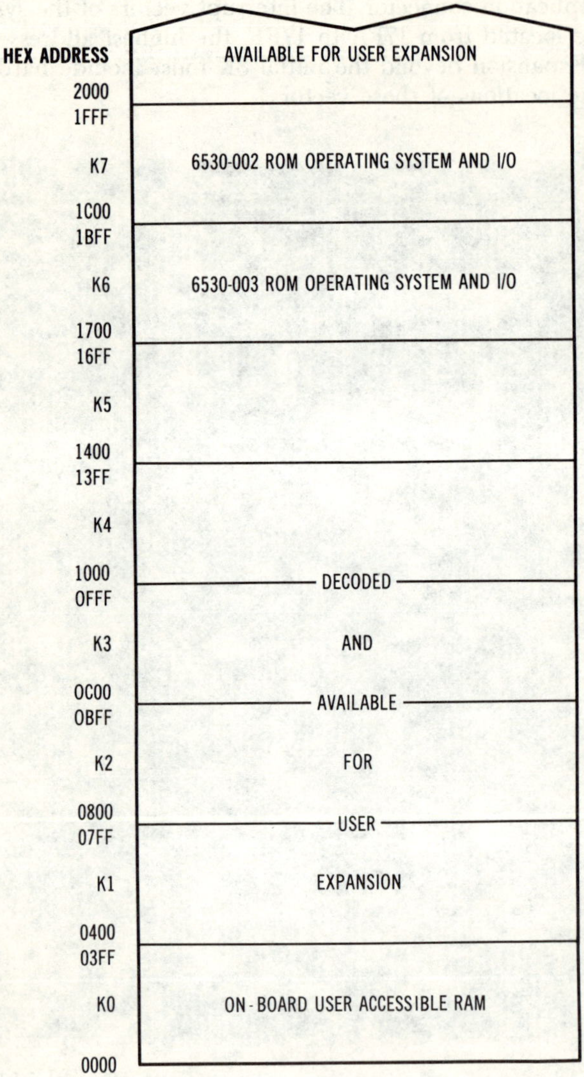

Courtesy MOS Technology, Inc.

Fig. 10-2. KIM-1 memory map.

KIM-1 EXPANSION-BOARD SIGNAL DEFINITIONS

The KIM-1 expansion board pinout designations are shown in Fig. 10-3.

Pin No.	Signal	Definition
8–15	DB0–DB7	Data bus—The bidirectional data bus lines carry information between the processor and all peripherals. These lines are pulled directly off the processor and are not buffered.
V	R/W	Read/write—The read/write line is used by the processor to dictate data flow on the data bus. When high, the processor is reading data from memory. When low, it is writing data to memory. The R/W line is buffered.
W	$\overline{\text{R/W}}$	Read/write—The $\overline{\text{R/W}}$ line is the buffered complement of the R/W line.

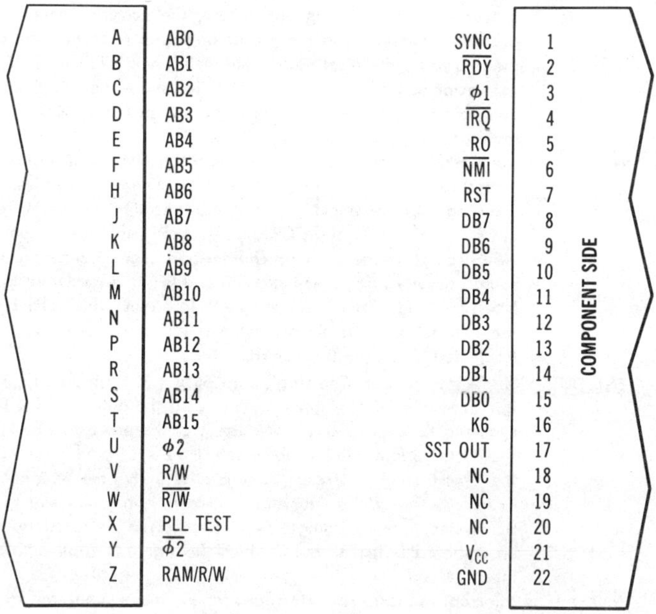

Fig. 10-3. KIM-1 expansion-connector pinout designations.

Pin No.	Signal	Definition
Z	RAM/R/W	RAM/R/W—The RAM/R/W line is the NANDed combination of the ϕ2 clock signal and the inverted read/write signal. Since all data transfers occur during the ϕ2 clock cycle, this line activates the RAM when valid data is present.
1	SYNC	Sync—The processor sets the sync line high during the ϕ1 clock cycle each time it reads an instruction from memory. The sync is pulled low at the end of that clock cycle. By observing the sync, external logic can use the $\overline{\text{RDY}}$ line to allow the processor to work with slow or dynamic memory or DMA controllers.
2	$\overline{\text{RDY}}$	Ready—The ready line is an input that causes the processor to suspend internal operations, allowing the processor to wait on slow memories or dynamic refresh cycles. The processor will respond to a falling edge on the ready line on any ϕ1 clock cycle except during a write operation. If the ready is pulled low during a write

Pin No.	Signal	Definition
		operation, the processor will ignore it until the $\phi 1$ clock of the first nonwrite operation that follows.
A–F, H, J–N, P, R–T	AB0–AB15	Address bus—The unbuffered address bus lines connect the processor and peripherals, thus providing the needed address.
16	K6	K6—The K6 line is the output of an address decoder that is pulled high when the processor puts an address between hex 1800 and 1BFF on the address bus. The K6 is used on the KIM-1 in conjunction with K5 (another select line) to enable a 6530 ROM that holds some of the system monitor.
7	$\overline{RST}$	Reset—The reset line is directly connected to the reset line of the processor. If this line is pulled low, the current operation of the processor is halted and the processor begins executing the program pointed to by the reset vector. Normally, hex FFFE and FFFF, the reset vector of the KIM-1, is hardware located at hex 17FC and 17FD. This points the processor to the beginning of the ROM-stored system monitor.
6	$\overline{NMI}$	Nonmaskable interrupt—The nonmaskable interrupt input is connected to the $\overline{NMI}$ line of the processor. When the processor detects a falling edge on the $\overline{NMI}$, it stops execution of the current program, saves the PC and CCR on the stack, jams the contents of the $\overline{NMI}$ vector into the program counter and begins execution. The $\overline{NMI}$ vector, usually located at hex FFFC and FFFD, is hardware located at hex 17FA and 17FB. These are RAM locations which can be set by the user to point to his or her interrupt service routine. Only the reset input has priority over the $\overline{NMI}$.
17	SST OUT	Single-step output—The single-step output line allows single instruction execution of a program by applying the output of the NANDed sync and K7 signals to the NMI input of the processor. By NANDing the K7 (an address line used to select the system monitor ROM) and the sync line, an interrupt is generated each time an instruction is executed. The NMI vector must point to a register-save routine in the system monitor. Single-step operation is defeated by opening an on-board switch through which the signal is applied to the NMI input.
4	$\overline{IRQ}$	Interrupt request—The interrupt request line is connected to the $\overline{IRQ}$ input of the processor. When the $\overline{IRQ}$ is pulled low, the processor checks the "I" status flag. If it is low, execution of the current program is halted, the PC and CCR are saved on the stack, the contents of the $\overline{IRQ}$ vector is jammed into the program counter and execution is continued. If the "I" flag is high, the processor will ignore the $\overline{IRQ}$ request until the flag is reset. The $\overline{IRQ}$ vector, normally located at FFFA and FFFB, is hardware located at hex RAM locations 17FE and 17FF. It can be set by the user. Both the reset and $\overline{NMI}$ inputs have priority over the $\overline{IRQ}$.
3	$\phi 1$	$\phi 1$—The $\phi 1$ clock signal is taken from the processor's $\phi 1$ output. The $\phi 1$ signal is generated internally by the processor and serves with the $\phi 2$ signal as the clock of the system.
U	$\phi 2$	$\phi 2$—The $\phi 2$ clock output signal is the buffered output of the $\phi 2$ line of the processor. Generated internally by the processor, the $\phi 2$ signal serves with the $\phi 1$ signal as the system clock. Data is transferred between the processor and peripherals during the $\phi 2$ signal.
Y	$\overline{\phi 2}$	$\overline{\phi 2}$—The $\overline{\phi 2}$ signal is the $\phi 2$ signal inverted.

Pin No.	Signal	Definition
5	RO	Reset overflow—The RO line connects to the RO input of the processor. A negative edge on this pin sets the overflow status flag of the processor.
X	PLL TEST	Phase-locked loop test—The phase-locked loop test output is taken from the audio cassette interface. It allows adjustment of the audio output frequency.
21	V_{CC}	V_{CC}—The V_{CC} input provides power to the KIM-1. It requires +5 Vdc at approximately 1.0 ampere.
22	GND	Ground—The ground line is the power and signal return supply.
18–20	NC	NC—These pins are not connected.

KIM-4 SIGNAL DEFINITIONS

Most of the application-connector signals are duplicated on the MOS Technology, Inc., KIM-4 shown in Figs. 10-4 and 10-5. As an expansion motherboard that plugs directly onto the KIM-1, the KIM-4 buffers all signals and provides connection slots for six additional boards, each of which uses KIM-4 bus. The KIM-4 includes address decode circuitry that activates the KIM-1 memory (when called). The pinout designations for the KIM-4 board are shown in Fig. 10-6.

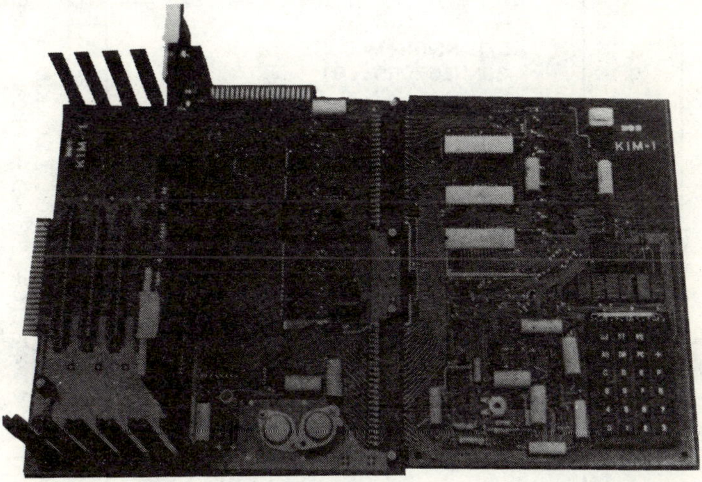

Courtesy MOS Technology, Inc.

Fig. 10-4. The KIM-4 plugged onto the KIM-1.

Pin No.	Signal	Definitions
8–15	BDB0–BDB7	Data lines—This is the buffered bidirectional data bus line.
W	BR/W	Read/write—This is the buffered read/write line.
2	BSYNC	BSYNC—This is the buffered sync line.
3	$\overline{BRDY}$	Ready—This is the buffered ready line.

Pin No.	Signal	Definitions
B–F, H, J–N, P, R–T, U	BAB0–BAB15	Address bus—This is the buffered address bus.
7	$\overline{\text{BRST}}$	Reset—This is the buffered reset line.
6	$\overline{\text{BNMI}}$	Nonmaskable interrupt—This is the buffered nonmaskable interrupt line.
4	$\overline{\text{BIRQ}}$	Interrupt request—This is the buffered interrupt request line.
V	Bϕ2	ϕ2—The buffered ϕ2 clock line.
X	Bϕ2	ϕ2—The buffered ϕ2 clock line inverted.
18	$\overline{\text{DMA}}$	Direct memory access—The direct memory access line is an input which, when pulled low, causes the data and address buses and the read/write line to float. Floating these lines means their buffers enter a high-impedance state which effectively disconnects the expansion bus from the processor. An external device such as a disk controller can then use the bus to load from or dump to the memory.

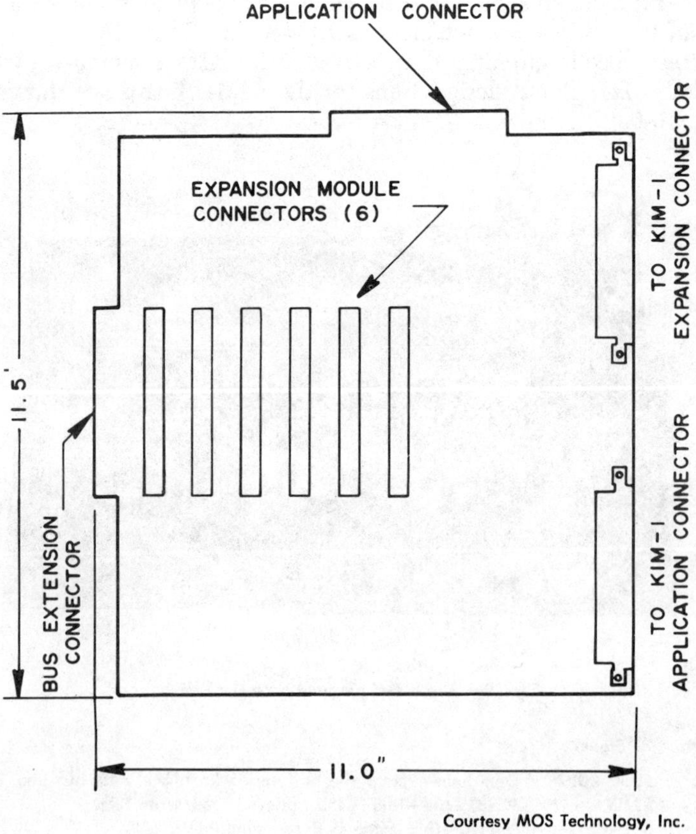

Courtesy MOS Technology, Inc.

Fig. 10-5. Diagram of the KIM-4 motherboard.

Pin No.	Signal	Definitions
21, Y	+5 V	+5 V—The +5-Vdc line provides a regulated +5 Vdc for the KIM-1.
19, 20	+8 V	+8 V—The +8-Vdc line provides unregulated power to each of the expansion boards which must contain regulation circuitry. The +8-volt supply should be capable of producing 15 amperes.
17	+15 V	+15 V—The +15-V line is an input that feeds a +12-volt regulator. The regulated +12 volts is passed to the KIM-1 for use in the audio interface circuitry.
5	−15 V	−15 V—The −15-V line is an unregulated input that is passed to each of the expansion boards where (if needed) it can be regulated.
1, 22, A, Z	GND	Ground—The ground line is the common return line for all system power supplies and signal lines.
16	BD SEL	This is an unused line common to all KIM-4 boards.

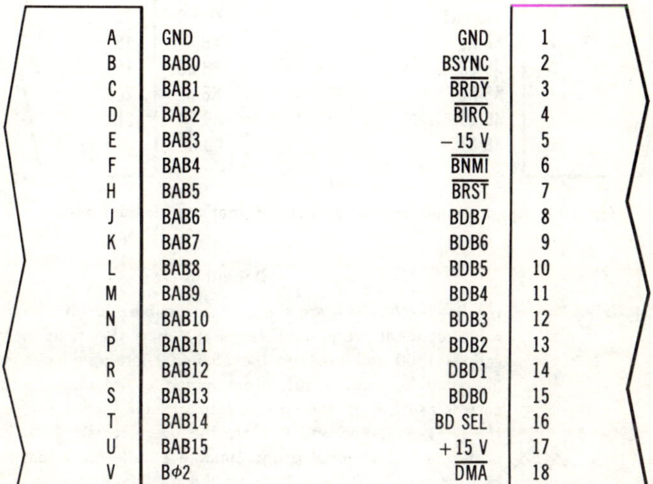

Fig. 10-6. Bus pinout designations for the KIM-4 board.

KIM-1 APPLICATION-CONNECTOR SIGNAL DEFINITIONS

Expansion of the KIM-1 is also possible through its application connector which provides i/o lines. The KIM-1 application connector is duplicated pin for pin on the KIM-4 motherboard except as noted in the signal definitions. The application-connector pinout designations for the KIM-1 are shown in Fig. 10-7.

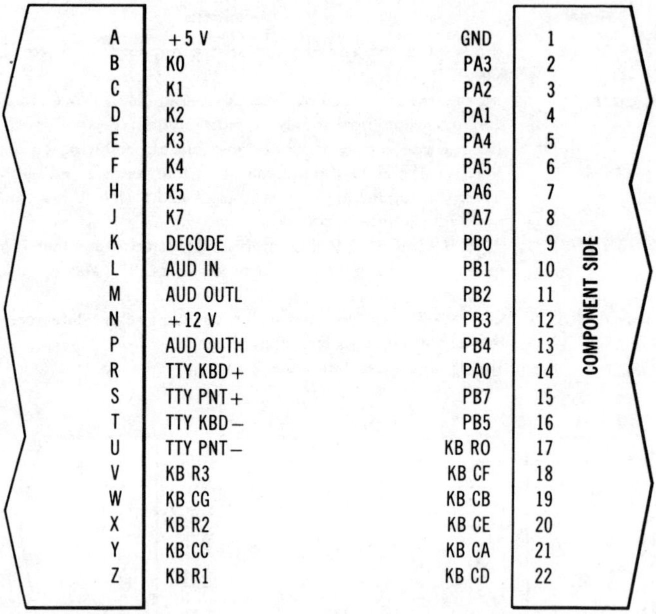

Fig. 10-7. Application-connector pinout designations for the KIM-1.

Pin No.	Signal	Definition
2–8, 14	PA0–PA7	I/o lines—PA0–PA7 are the user accessible i/o lines of port A, a bidirectional programmable parallel port that rests at hex locations 1700 and 1701 on the 6530-003 integrated circuit. Each line may be individually programmed as an input or output; each is capable of driving one standard TTL load.
9–13, 15, 16	PB0–PB5, PB7	I/o lines—PB0–PB5 and PB7 are the user accessible i/o lines of port B, a bidirectional programmable parallel port that rests at hex locations 1702 and 1703 on the 6530-003 integrated circuit. Each line is TTL compatible. Only seven lines are available for i/o since the line normally used for PB6 has been dedicated as a chip select.
18–22, W, Y	KB CA–KB CG	Keyboard column—These lines are i/o lines PA0–PA6 of port A on the 6530-002 integrated circuit. The KIM-1 uses these lines in the strobed hex keyboard and display. They are available to allow a remote keyboard. They can, however, also be used to i/o lines during program execution.
17, V, X, Z	KB R0–KB R3	Keyboard row—These lines are the output of a decimal decoder used by the KIM-1 (in conjunction with the keyboard column lines) to scan the hex keyboard. They are provided to allow a remote keyboard.
L	AUD IN	Audio input—The audio input line that is used to connect the earphone jack of an audio tape recorder to the on-board audio interface.
M, P	AUD OUT (Hi and Lo)	Audio output—The audio output lines connect the on-board audio interface to the microphone input on an audio tape recorder.

Pin No.	Signal	Definition
		The low output is normally used although recorders that have an auxiliary input may operate with the high output.
R, T	TTY KY (+ and −)	Teletypewriter keyboard—These lines provide a 20-mA current loop that can be driven by a teleprinter keyboard.
S, U	TTY PTR (+ and −)	Teletypewriter printer—These lines provide a 20-mA current loop that can drive a teleprinter print mechanism.
B–F, H, J	K0–K7	K0–K7—These lines are the outputs of a decimal decoder. They serve as chip select lines for the first 8 kilobytes of memory. See Fig. 10-2 for memory-map details.
K	DECODE	Decode—The decode line is an input to the K0–K7 decoder. It must be held low to enable the decoder and to use memory on the KIM-1. Expansion beyond 8 kilobytes requires that the external logic control the decode line to choose on- or off-board memory.
A	V_{CC}	V_{CC}—This is a +5-Vdc regulated input.
1	GND	Ground—The ground line is the V_{CC} return.
N	+12 V	+12 V—The +12-volt line.

DISCUSSION

KIM-1 expansion is often performed with S-100 boards since the KIM-compatible boards are scarce and expensive while the S-100 boards are plentiful and inexpensive. A commercially available interface board is discussed in Chapter 24.

CHAPTER 11

The Apple II/III Bus

The Apple II microcomputer, shown in Fig. 11-1, is a single-board system that includes a processor, operating system, and BASIC in ROM, RAM, keyboard, color video-display generator, cassette interface, power supply, speaker, and game paddles. In addition to the speaker and game paddles, the Apple II comes equipped with demonstration tapes, instruction manual, and your choice of either 8K or 16K of memory. Packaged in a compact plastic case, it has been a popular system.

Memory expansion is extremely simple on the Apple microcomputer. Three rows of sockets are provided that will support either 4- or 16-kilobit dynamic RAMs. The system can be expanded by simply filling the existing sockets with chips or by replacing 4-kilobit chips with 16-kilobit chips. A maximum of 48 kilobytes of RAM is possible using these sockets.

A motherboard area (containing eight sockets) is also provided. It is located at the back of the main circuit board as shown in Fig. 11-2. All signals necessary to add memory or peripherals are available at these connectors.

A variety of expansion boards, like the one shown in Fig. 11-3, are offered by Apple Computer, Inc. The expansion boards are 2¾ by 7 inches (6.9 by 17.7 cm) in size and use a double-sided, 50-pin (25 each side), edge-card connector with 0.1-inch (0.25-cm) spacing. The connectors are offset as shown in Fig. 11-3, providing positive indexing for the boards. The expansion board shown in Fig. 11-3 is Apple's Model A2B0003X Communications Card. It permits the Apple II to serve as a terminal in a time-sharing network or within a network of two or more Apple computers that have access to and control of other Apple II computers.

Courtesy Apple Computer, Inc.

Fig. 11-1. The Apple II microcomputer.

APPLE SIGNAL DEFINITIONS

The Apple bus pinout designations are shown in Fig. 11-4.

Pin No.	Signal	Definition
42–49	D0–D7	Data lines—These are bidirectional, buffered data lines of the processor. They carry information into and out of the processor.
18	R/$\overline{W}$	Read/write—The buffered read/write line of the processor is used to control the direction of data flow on the data lines. When high, a read operation will occur, when low, a write operation will occur.
21	$\overline{RDY}$	Ready—The ready line is a processor input that, when pulled low during $\phi 1$, will halt the processor during the next $\phi 1$ cycle. When it is pulled high, execution will continue in the fol-

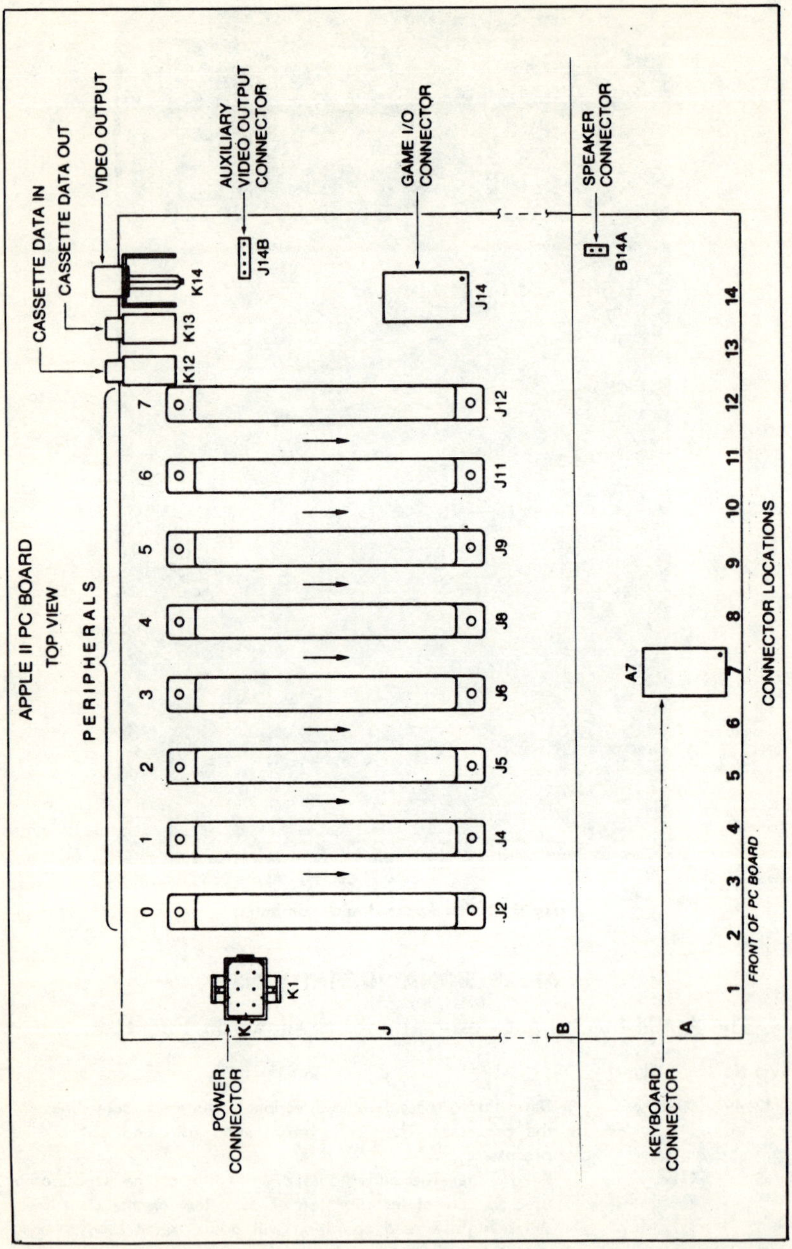

Courtesy Apple Computer, Inc.

Fig. 11-2. The Apple II pc board showing the "motherboard area" located at the rear of the main board.

Courtesy Apple Computer, Inc.

Fig. 11-3. A serial-communications expansion board.

Pin No.	Signal	Definition
		lowing $\phi 2$. Slow memory can use the ready line to gain the extra time needed to access data. The ready signal will not halt the processor during a write operation, but in the $\phi 1$ that follows.
2–17	A0–A15	Address lines—The buffered address lines of the processor carry the location of the desired data from the processor to the peripherals.
20	$\overline{\text{I/O STROBE}}$	I/o strobe—The i/o strobe is an active-low output. It is pulled low on all connectors when a hex address between C800 and CFFF is called by the processor. It can be used as a chip select signal. (See the memory map of Fig. 11-5.)
41	$\overline{\text{DEVICE SELECT}}$	Device select—The device select line is an active-low output that is pulled low on only one connector at a time in response to predetermined addresses. As shown in Fig. 11-5, each connector is assigned 16 unique addresses that are associated with the device select. When an address assigned to a connector is called, the device select line is pulled low on only that connector. This line can be used as a chip select signal.
1	$\overline{\text{I/O SELECT}}$	I/o select—The i/o select is an active-low output that, like the device select, is pulled low on only one connector at a time, in response to a specific group of addresses. As shown in Fig. 11-5, each connector, except No. 0, is assigned 256 unique addresses that are associated with the i/o select. When an

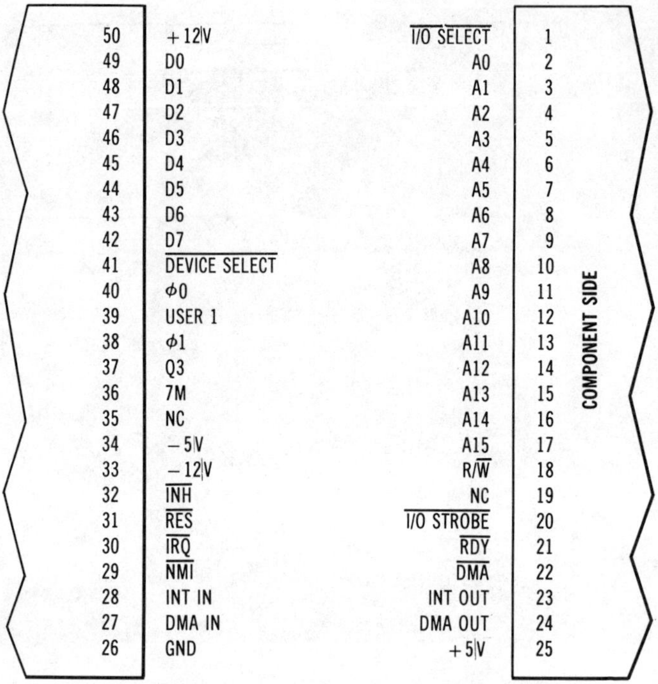

Fig. 11-4. The Apple bus pinout designations.

Pin No.	Signal	Definition
		address assigned to a connector is called, the i/o select line is pulled low on only that connector. This line can be used as a chip select signal.
32	$\overline{INH}$	Inhibit—The inhibit line is an input that, when pulled low, disables all on-board ROMs, thus allowing the use of customized operating system software.
31	$\overline{RES}$	Reset—The reset line is an input to the active-low reset of the processor. Pulling this line low will cause the processor to cease its current operation and return to the system monitor.
29	$\overline{NMI}$	Nonmaskable interrupt—The $\overline{NMI}$ line is an input to the active low $\overline{NMI}$ of the processor. When pulled low, the processor will finish the current instruction, push the program counter and status register onto the stack, load the contents of the $\overline{NMI}$ vector into the program counter and, then, begin executing the $\overline{NMI}$ service routine.
30	$\overline{IRQ}$	Interrupt request—The $\overline{IRQ}$ line is an input to the active-low $\overline{IRQ}$ of the processor. When pulled low, the processor will finish the current instruction and check the "I" status flag. If it is reset, the program counter and status register will be pushed onto the stack, the contents of the $\overline{IRQ}$ vector will be loaded into the program counter and execution of the $\overline{IRQ}$ service routine will begin. If the "I" flag is set, the processor will ignore an $\overline{IRQ}$ input until it is reset.

HEX ADDRESS	PIN FUNCTION
CFFF C800	THE I/O STROBE IS PULLED LOW ON ALL CONNECTORS.
C7FF C700	THE I/O SELECT IS PULLED LOW ON CONNECTOR 7.
C6FF C600	THE I/O SELECT IS PULLED LOW ON CONNECTOR 6.
C5FF C500	THE I/O SELECT IS PULLED LOW ON CONNECTOR 5.
C4FF C400	THE I/O SELECT IS PULLED LOW ON CONNECTOR 4.
C3FF C300	THE I/O SELECT IS PULLED LOW ON CONNECTOR 3.
C2FF C200	THE I/O SELECT IS PULLED LOW ON CONNECTOR 2.
C1FF C100	THE I/O SELECT IS PULLED LOW ON CONNECTOR 1.
C0FF C0F0	THE DEVICE SELECT IS PULLED LOW ON CONNECTOR 7.
C0EF C0E0	THE DEVICE SELECT IS PULLED LOW ON CONNECTOR 6.
C0DF C0D0	THE DEVICE SELECT IS PULLED LOW ON CONNECTOR 5.
C0CF C0C0	THE DEVICE SELECT IS PULLED LOW ON CONNECTOR 4.
C0BF C0B0	THE DEVICE SELECT IS PULLED LOW ON CONNECTOR 3.
C0AF C0A0	THE DEVICE SELECT IS PULLED LOW ON CONNECTOR 2.
C09F C090	THE DEVICE SELECT IS PULLED LOW ON CONNECTOR 1.
C08F C080	THE DEVICE SELECT IS PULLED LOW ON CONNECTOR 0.

Courtesy Apple Computer, Inc.

Fig. 11-5. Special i/o connector select addresses.

Pin No.	Signal	Definition
23, 28	INT IN, INT OUT	Interrupt input and interrupt output—The INT IN and INT OUT lines allow the implementation of a simple interrupt priority scheme in a system that has more than one source of interrupts. The IN OUT line of each connector runs to the INT IN of the next connector to form the simple "daisy chain" shown in Fig. 11-6. The status of these lines determines when a board can request service on the NMI or IRQ lines. When an interrupt is generated on a board, the board cannot signal the IRQ until it checks the status of its INT IN. If the INT IN of the board is low, a higher priority board is using the IRQ and the board must wait until the INT IN is returned high. If the INT IN is high, or is returned high, the board can use the IRQ but it must first pull its INT OUT low, thereby indicating a busy IRQ to boards with lower priority. Each board must pass a busy signal on to lower priority boards by pulling its INT OUT low when the INT IN of the board is pulled low. Using this scheme, the board in connector 0 has the highest priority and the board in position 7 has the lowest. Any board not using interrupts must pass the INT IN to the INT OUT with a jumper to prevent a break in the chain. All hardware needed to implement this system must be contained on the peripheral boards. It is important to note that use of the INT IN and INT OUT is completely optional. The NMI and IRQ are provided at each connector and can be used as desired with or without the use of the INT IN and INT OUT.
40	$\phi 0$	Clock phase 0—This signal is derived from the master oscillator and is used to drive the clock input of the processor. Using this signal, the processor produces the $\phi 1$ and $\phi 2$ signals for the system. This $\phi 0$ signal can be used the same as $\phi 2$ since they are functionally the same.

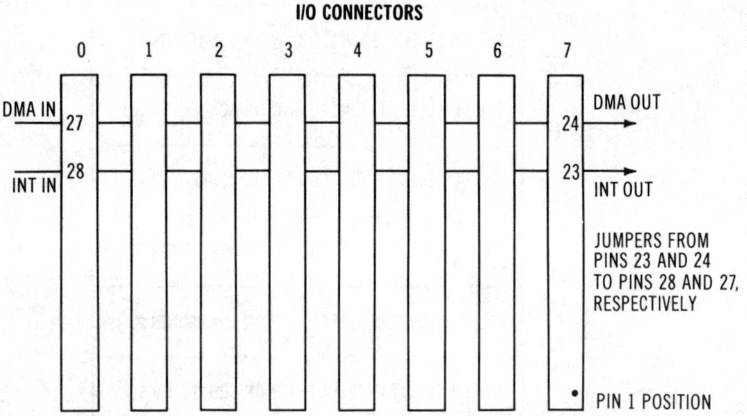

Fig. 11-6. The INT IN/INT OUT and DMA IN/DMA OUT "daisy chains" on the i/o connectors.

Pin No.	Signal	Definition
38	$\phi 1$	Clock phase 1—The $\phi 1$ signal is one output of the system clock generator of the processor.
36	7M	7-MHz clock—This 7-MHz signal is derived from the master oscillator and is available for use by peripherals.
37	Q3	Q3—The Q3 is a 2-MHz signal derived from the master oscillator. It is available for use by peripherals.
22	$\overline{\text{DMA}}$	Direct memory access—The $\overline{\text{DMA}}$ line can be used by peripherals to gain control of the buses of the system. Pulling this line low floats the address and read/write lines and inhibits the $\phi 0$ clock input, effectively stopping the processor and putting the data lines in the input mode. A peripheral can then use the buses for DMA operations. Since the 6502 is a dynamic device, the clock can only be stopped for a maximum of 1 ms before the internal register contents are lost. Therefore the $\overline{\text{DMA}}$ must be completed in less than 1 ms.
24, 27	DMA IN, DMA OUT	Direct memory access in and direct memory access out—The DMA IN and DMA OUT lines allow the implementation of a DMA priority scheme much like the interrupt priority that was obtained using the INT IN and INT OUT lines. The DMA and DMA OUT lines are connected (Fig. 11-6) to form a "daisy chain" which functions like the INT IN and INT OUT chain. Use of the DMA IN and DMA OUT lines is completely optional; the DMA functions without their use.
39	USER 1	User 1—The user line is an input to the address decoder that generates the i/o and ROM enable signals. Pulling this line low disables the i/o enable signals and all on-board ROM, thus allowing different implementation of the memory space between hex C000 and FFFF.
19, 35	NC	No connection—The NC contact is undefined and available for use.
25	+5 V	+5 V—The +5-Vdc line is a regulated output that can provide up to 500 mA.
50	+12 V	+12 V—The +12-Vdc line is a regulated output that can provide up to 250 mA.
34	−5 V	−5 V—The −5-Vdc line is a regulated output that can provide up to 200 mA.
33	−12 V	−12 V—The −12-Vdc line is a regulated output that can supply up to 200 mA.
26	GND	Ground—The ground line is the common return for the signal and power lines of the system.

CHAPTER 12

The PET Bus

Commodore Business Machines' PET, shown in Fig. 12-1, is one of the new breed of microcomputers. It is a complete system in a single package. The PET system provides CPU, RAM, system executive and BASIC in ROM, integral alphanumeric keyboard and crt display, cassette recorder and interface, programmable i/o lines,

Courtesy Commodore Business Machines, Inc.

Fig. 12-1. The PET system.

and instrumentation provisions in an attractive case. The user has only to plug the unit in and turn it on to get started in a BASIC operating system. Introduced in 1977, the PET has been very popular. Several expansion modules for the system are currently being offered.

Expansion of the 6502-based system is relatively easy. All of the expansion lines are buffered and the expansion memory is decoded into 4-kilobyte blocks. Expansion of the PET is performed via a 40-pin edge-card connector (with 0.100-inch, or 0.25-cm, spacing) that protrudes from the right side of the PET. All of the pins on the top side of the connector are grounded. No positive indexing is provided.

PET EXPANSION-CONNECTOR SIGNAL DEFINITIONS

The PET expansion-connector pinout designations are shown in Fig. 12-2.

Pin No.	Signal	Definition
1–79	—	All odd numbered pins are grounded.
66, 68, 70, 72, 74, 76, 78, 80	D0–D7	Data bus—The bidirectional data bus lines carry data between the processor and the peripherals.
60	R/W	Read/write—The read/write line is an output used by the processor to control direction of data movement on the data bus. When high, the processor is reading data; when low, it is writing data.
2, 4, 6, 8, 10, 12, 14, 16, 18, 20, 22, 24	A0–A11	Address bus—The address bus lines are outputs that are used by the processor, in conjunction with the select lines, to specify the location desired.
32, 34, 36, 38, 40, 42, 44, 46, 48, 50	SEL1–SEL7, SEL9–SELB	Select—The select lines are the decoded address outputs of the A12–A15 lines of the processor, that do not appear on the connector. SEL1–SELB are used to select, or enable, 4-kilobyte blocks of expansion memory or i/o. Select line 1 enables the 4-kilobyte block between hex 1000 and 1FFF, select line 2 enables addresses between hex 2000 and 2FFF, etc. The processor uses the select signals to indicate a specific 4-kilobyte block and the A0–A11 lines to indicate the desired byte in that block. The memory map for the PET is shown in Fig. 12-3. Dedicated and user-available areas are defined using the select lines.
54	$\overline{RES}$	Reset—The reset line is an unbuffered input to the reset line of the processor. Pulling this line low causes the processsor to load the program counter with the contents of the reset vector and to begin execution.
56	$\overline{IRQ}$	Interrupt request—The interrupt request line is an unbuffered input to the $\overline{IRQ}$ of the processor. When pulled low, the processor examines the "I" flag and, if reset, sets the "I" flag, stores the PC and status register on the stack, loads the PC with the contents of the $\overline{IRQ}$ vector and begins execution of the $\overline{IRQ}$ service routine.

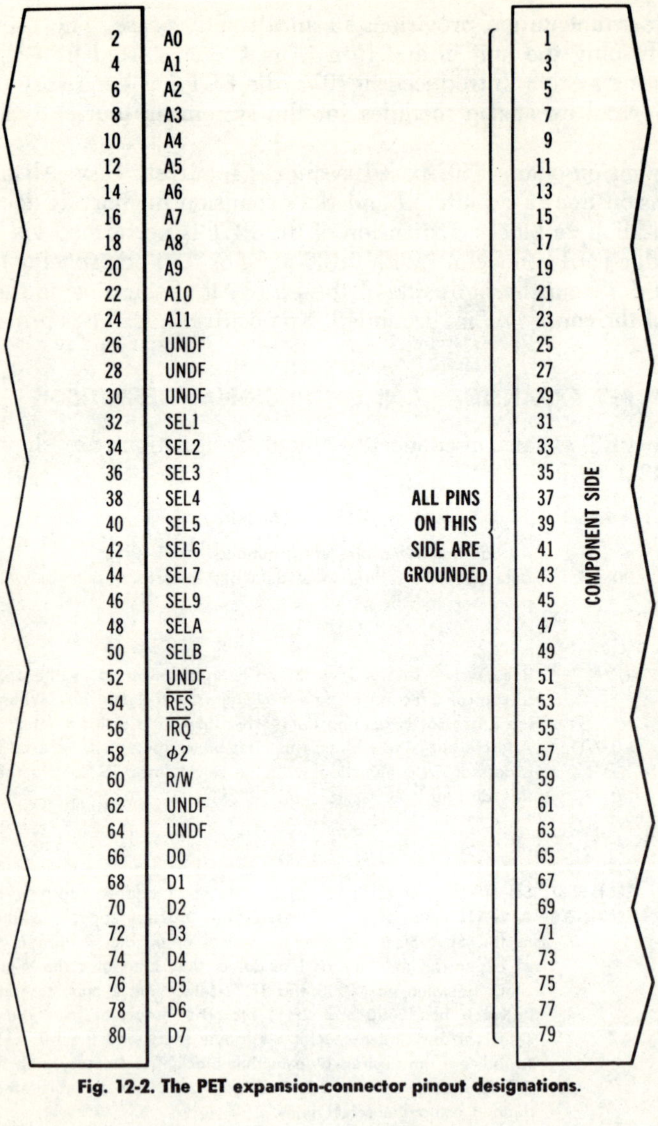

Fig. 12-2. The PET expansion-connector pinout designations.

Pin No.	Signal	Definition
		If the "I" flag is set when the $\overline{IRQ}$ is pulled low, the processor will ignore the $\overline{IRQ}$ until the "I" is reset.
58	$\phi 2$	$\phi 2$—The $\phi 2$ line is the output of the system clock and can be used as a chip enable since all data movement occurs during $\phi 2$.
26, 28, 30, 52, 62, 64	UNDF	Undefined—These lines are not assigned a function by the PET and can be used externally for user desired functions.

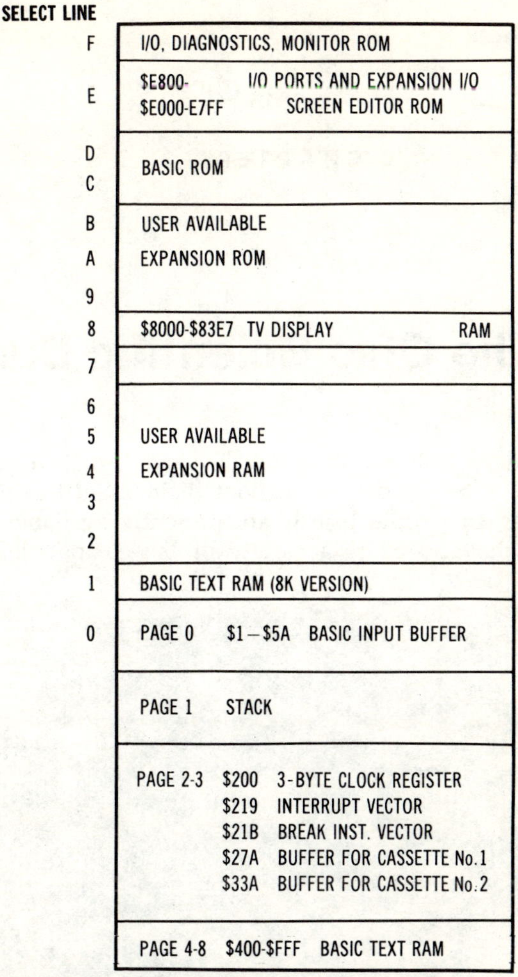

Courtesy Commodore Business Machines, Inc.

Fig. 12-3. The PET memory map.

Several companies are producing PET-to-S-100 interface boards that allow PET owners to use the many available S-100 boards. See Chapter 25 for further details.

The PET also contains a connector with IEEE 488 signals. This edge connector must be cabled to a standard IEEE 488 connector to be used with other 488 devices. Implementation of the IEEE 488 on the PET is accomplished with software drivers. The IEEE 488 is defined in Chapter 22.

CHAPTER 13

The Ohio Scientific Bus

The standard 48-line system bus developed in 1975 by Ohio Scientific, Inc., is used today throughout their broad line of products. Over 40 bus-compatible boards are presently available from Ohio Scientific, who also offers a variety of bus-compatible microcom-

Courtesy Ohio Scientific, Inc.

Fig. 13-1. The Challenger III microcomputer.

Courtesy Ohio Scientific, Inc.

Fig. 13-2. The Challenger C2-4P microcomputer.

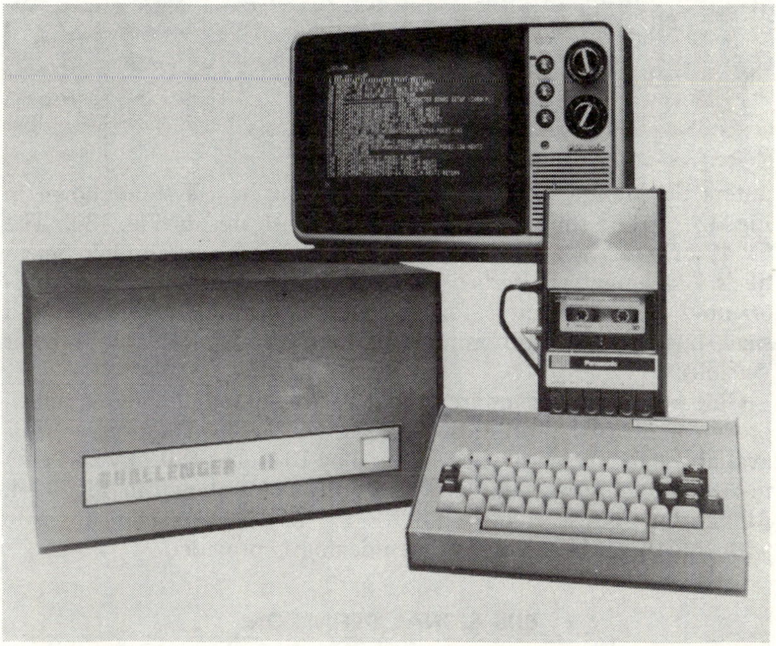

Courtesy Ohio Scientific, Inc.

Fig. 13-3. The Challenger C2-8P microcomputer.

Courtesy Ohio Scientific, Inc.

Fig. 13-4. The OS-48 motherboard.

puters that range from the expandable business system shown in Fig. 13-1 to the simple turn-key system illustrated in Fig. 13-2. The C2-4P (Fig. 13-2) and the C2-8P (Fig. 13-3) are part of Ohio Scientific's Challenger IIP series. These are ultra-high-performance bus-oriented microcomputers for personal, educational, research, and small business use. Both microcomputers are also available in color versions.

Ohio Scientific systems are primarily 6502 based although a multi-processor board featuring 6502, 6800, and Z-80 microprocessors is available. Ohio Scientific boards are 8 by 10 inches (20.3 by 25.4 cm) in size and plug into the backplane connector shown in Fig. 13-4. All the boards use four 12-pin KK-156 Molex connectors to mate with the 48-pin bus. No positive indexing is provided.

BUS SIGNAL DEFINITIONS

The pinout designations of the Ohio Scientific bus are shown in Fig. 13-5.

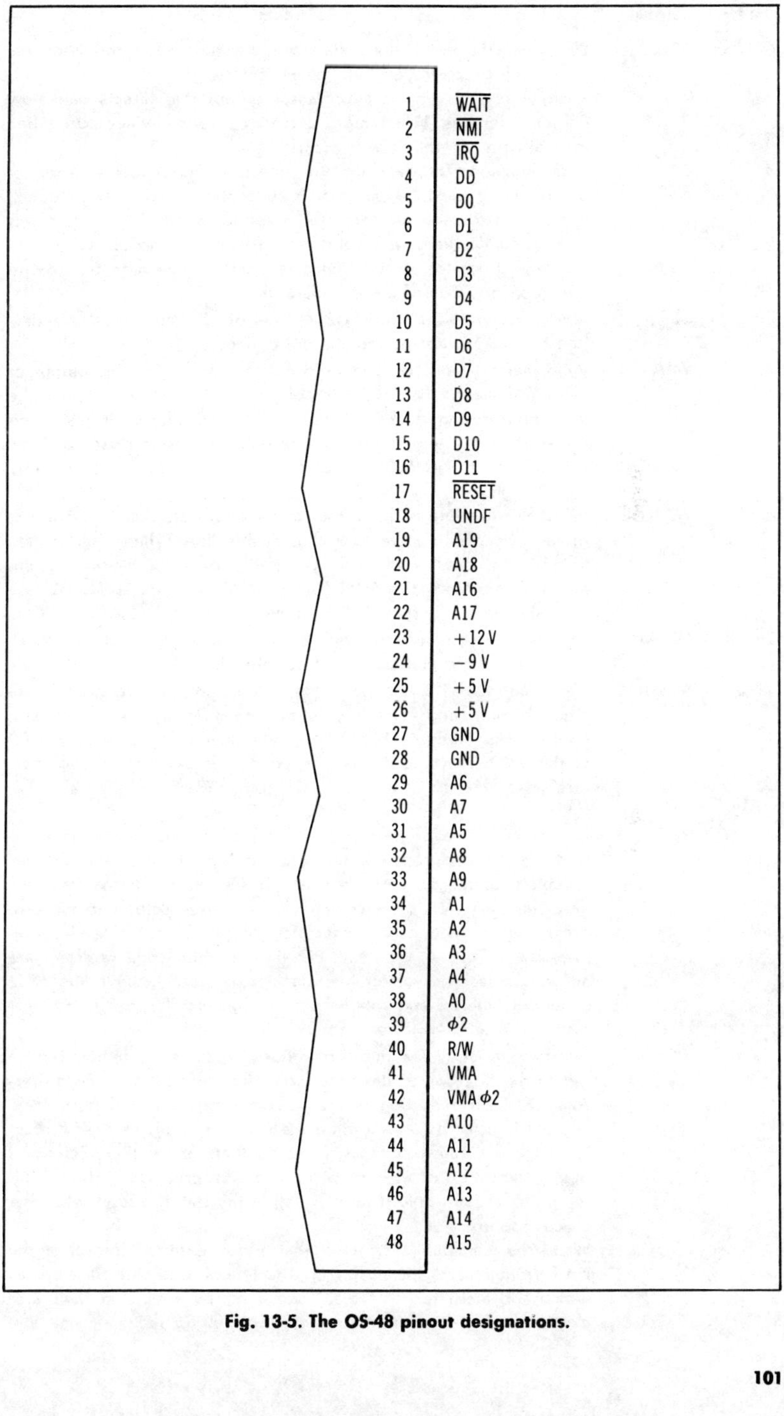

Fig. 13-5. The OS-48 pinout designations.

Pin No.	Signal	Definition
5–12	D0–D7	Data line—The bidirectional data bus which carries information between the processor and all the peripherals.
40	R/W	Read/write—This line is a processor output that directs data flow on the data bus. When high, a read operation will occur; when low, a write operation will occur.
4	DD	Data direction—The data direction line is an input to the processor board that is used by peripherals to control its data bus buffers. When a peripheral board, after responding to the address bus (R/W) and $\phi 2$ lines, has valid data for or can accept valid data from the processor, it pulls the data direction line high for a write operation or, low for a read operation.
39	$\phi 2$	Clock phase 2—The clock $\phi 2$ output of the processor. All data transfers take place during the $\phi 2$ period.
41	VMA	Valid memory address—The valid memory address is the output of the VMA line of the 6800 microprocessor. The processor pulls the VMA high during clock $\phi 1$, after the desired address is placed on the address bus and is settled. The 6502 processor does not have a VMA line so, in 6502 systems, line number 41 is permanently pulled high.
42	VMA $\phi 2$	VMA $\phi 2$—This line is the output of an AND gate that is fed to peripheral boards to serve as a chip enable line. When high, a data transfer can occur with little possibility of error caused by unsettled lines. Since the 6502 has no VMA line, the VMA $\phi 2$ line on 6502 systems is the $\phi 2$ line of the processor.
29–38, 43–48	A0–A15	Address bus—The address bus lines are processor outputs which carry the desired address to the peripherals.
19–22	A16–A19	Upper address—These upper address lines are used to provide additional addressing capability in systems that are expanded with floppy disks. Signals A16–A19 are not inherent to either the 6502 or the 6800. They are derived by using four 6820 peripheral interface adapter lines configured as outputs. They are software controlled.
2	$\overline{NMI}$	Nonmaskable interrupt—This is the unbuffered nonmaskable interrupt input to the processor. When pulled low, the processor finishes execution of the current instruction, pushes some internal registers onto the stack, loads the contents of the $\overline{NMI}$ vector into the program counter, and begins execution of the interrupt service routine. The ACCA, ACCB, INX, PC and condition-code registers are all automatically saved on the stack in the 6800, while the 6502 saves only the PC and condition-code registers. The processor must respond to an $\overline{NMI}$ request for service.
3	$\overline{IRQ}$	Interrupt request—The interrupt request input is the unbuffered input to the $\overline{IRQ}$ line of the processor. When pulled low, the processor will finish execution of the current instruction and push some internal registers (see NMI) onto the stack. It then sets the "I" flag, loads the program counter with the contents of the $\overline{IRQ}$ vector and begins execution of the service routine. The processor will respond to an $\overline{IRQ}$ request only if the "I" flag in the status is reset when the request occurs.
1	$\overline{WAIT}$	Wait—The wait line is an input that allows external control of the clock frequency of the processor. The line is used to switch diodes which add resistance to the RC circuit of the astable multivibrator clock, thereby changing the frequency. Pulling the wait line low

Pin No.	Signal	Definition
		slows the clock to about 500 kHz to allow easy interface with slow memory and peripherals. The wait line will not be active in systems which are configured for a single clock speed.
17	$\overline{\text{RESET}}$	Reset—The reset line is an optional line that may, in some systems, be used to reset peripheral interface adapters.
13–16	D8–D11	Upper data—Optional data bus lines that may be used if a 12-bit processor (the Intersil 6100) is added to the system. They are not inherent to either the 6502 or the 6800 or their systems.
18	UNDF	Undefined—The undefined line has no dedicated use and may be used to perform special or custom functions.
25, 26	+5 V	+5 V—This line is a power input that requires +5 Vdc at 2.0 amperes regulated voltage.
23	+12 V	+12 V—This line is a power input that requires +12 Vdc at 0.5-ampere regulated voltage.
24	−9 V	−9 V—This line is a power input that requires −9 Vdc at 0.5-ampere regulated voltage.
27, 28	GND	Ground—The ground line is the common return for all system power supplies and signal lines.

All lines except the $\overline{\text{NMI}}$, $\overline{\text{IRQ}}$, and $\overline{\text{RESET}}$ lines are buffered. It should be noted that the processor reset line is not on the system bus. It is, instead, available on an auxiliary connector that is located on the CPU board.

AUXILIARY-CONNECTOR SIGNAL DEFINITIONS

Many of the systems offered by Ohio Scientific are based on their Model 500 CPU board, which is the nearer board shown in Fig. 13-4. Since it is so widely used, its auxiliary connector (AX) will be defined. The auxiliary connector accepts a KK-156 Molex connector, which is located in the upper left-hand corner of the Model 500 board (Fig. 13-4). The pinout designations of the auxiliary connector are shown in Fig. 13-6.

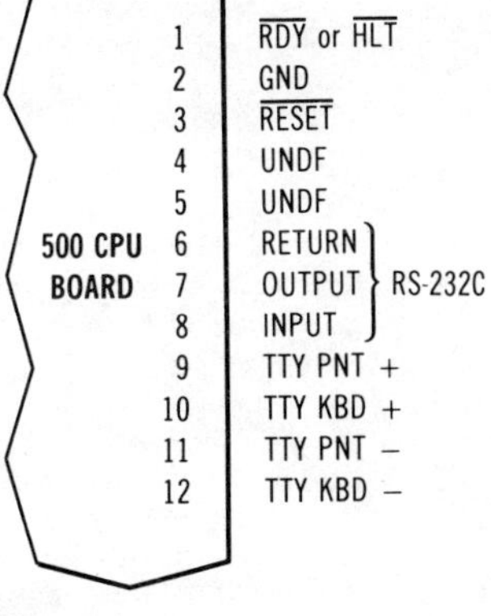

Fig. 13-6. Pinout designations for the auxiliary connector.

Pin No.	Signal	Definition
1	$\overline{\text{RDY}}$ or $\overline{\text{HLT}}$	Ready or halt—This line is an unbuffered input to the processor. Pulling this line low causes the processor to suspend operations and to float the buses until it is returned high. The ready line is available on the 6502 microprocessor, while the halt line is available on the 6800 microprocessor.
2	GND	Ground—This line is the return for system power supplies and signal lines.
3	$\overline{\text{RESET}}$	Reset—The reset line is the unbuffered input to the reset input of the processor. Pulling the reset low causes execution to begin at the location pointed to by the reset vector. A momentary contact push-button switch can be connected to pins 2 and 3 to perform the reset.
4, 5	UNDF	Undefined—Lines 4 and 5 are unused and allow special functions desired by the user.
6–8	RET, OUT, IN	RS-232C—Lines 6, 7, and 8 provide serial interface to the CPU board using RS-232C signals. Line 7 is the output, line 8 is the input, and line 6 is the common return line.
9, 11	TTY PNT (+ and −)	Teletypewriter printer—These lines provide serial printer interface to the CPU board using a 20-mA loop.
10, 12	TTY KBD (+ and −)	Teletypewriter keyboard—These lines provide serial keyboard interface to the CPU board using a 20-mA loop.

CHAPTER 14

The Atari Bus

Atari, Inc., drawing on their vast video-game experience, entered the personal computer field with the introduction of the Atari 400 and Atari 800 personal computer systems shown in Fig. 14-1. The systems are 6502 based and include 8 or 16 kilobytes of RAM, a 10-kilobyte operating system, a 16-color video output, a cassette interface, and an ASCII keyboard. Both systems also feature sockets, one in the 400 and two in the 800, that accept ROM-based program cartridges. These sockets, located in the computer consoles, allow the immediate execution of a variety of educational, game, and financial programs as well as the use of BASIC and assembly languages. Expansion of the systems can be accomplished by adding peripheral devices or by adding RAM.

Adding peripherals to either the 400 or the 800 involves use of a built-in asynchronous serial i/o bus. The i/o bus connector is located on the right side of the console and can interface a variety of peripherals, including cassette recorders, disk drives, printers, modems, and a serial/parallel expansion interface. The serial port pinout designations are shown in Fig. 14-2.

ATARI BUS SIGNAL DEFINITIONS

Pin No.	Signal	Definition
1	CLOCK INPUT	Clock input—The clock input line is currently unused but may be used in a future synchronous serial bus scheme.
2	CLOCK OUTPUT	Clock output—The clock output line is a processor system output that is used by peripherals to latch data from the data out line. Data is placed on the data out line during the rising edge of the clock output signal and is stable and can be latched on the falling edge of the clock output signal.

Courtesy Atari, Inc.

Fig. 14-1. The Atari 400 and 800 computer systems.

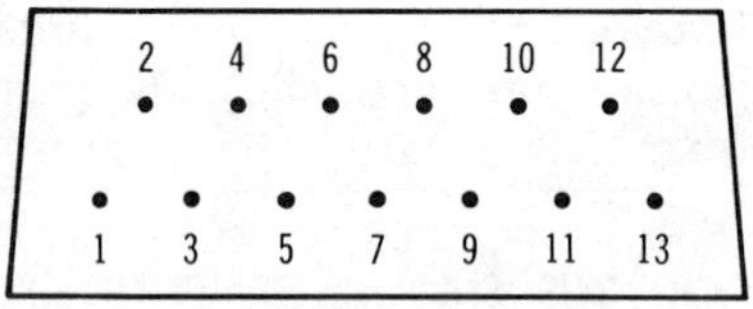

1 CLOCK INPUT
2 CLOCK OUTPUT
3 DATA INPUT
4 GROUND
5 DATA OUTPUT
6 GROUND
7 $\overline{\text{COMMAND}}$
8 MOTOR CONTROL
9 $\overline{\text{PROCEED}}$
10 +5/READY
11 AUDIO INPUT
12 +12V
13 $\overline{\text{INTERRUPT}}$

Fig. 14-2. Atari serial i/o port pinout.

Pin No.	Signal	Definition
3	DATA INPUT	Data input—The data input line is a processor system input used by peripherals to transmit data and status codes into the processor at a rate of 19,200 baud.
4	GROUND	Ground—The signal and power ground return line.
5	DATA OUTPUT	Data output—The data output line is a processor system output used by the processor to transmit commands and data to peripherals on the bus. Data on this line is stable and can be read on the falling edge of the clock output signal.
6	GROUND	Ground—The signal and power ground return line.
7	$\overline{\text{COMMAND}}$	Command—The command line is an active-low processor system output used to indicate to peripherals on the bus that the following 5 bytes are to be interpreted as a command frame.
8	MOTOR CONTROL	Motor control—The motor control line is a processor system output used to start and stop the cassette recorder/player's motor.
9	$\overline{\text{PROCEED}}$	Proceed—The proceed line is at present an active-low processor input that is connected to the 6502's $\overline{\text{IRQ}}$ line. When pulled low by a peripheral the processor may cease execution, store the program counter, status, and A registers on the stack and begin execution of the service routine pointed to by the vector at location 0202.
10	+5/READY	+5/ready—The +5/ready line is a processor system output that is pulled high by the system to indicate to peripherals that the system is powered up and ready. The +5/ready is also a processor input connected to the 6502's $\overline{\text{IRQ}}$ line that, when pulled low by a peripheral, may cause the 6502 to push the program counter and status and A registers onto the stack and begins execution of the service routine pointed to by the vector at location 020C.
11	AUDIO INPUT	Audio input—The audio input line is a processor system input from the cassette recorder. The cassette scheme used by Atari permits the recording of an audio track in addition to the digital data track. The output of the audio track is fed through the system to the tv set's audio section and speaker.
12	+12V	+12 volts—A limited-current +12-volt dc supply line.
13	$\overline{\text{INTERRUPT}}$	Interrupt—The interrupt line is an active-low processor input that is connected to the 6502's $\overline{\text{IRQ}}$ line. When a peripheral pulls the interrupt line low the processor will cease execution, push the program counter and status register onto the stack, and begin execution of the service routine pointed to by the vector at 0204.

DISCUSSION

Communication on the serial bus is initiated by the processor. All exchanges begin when the processor, on direction from the serial i/o firmware, composes a command frame of 5 bytes. The first byte is the address, or ID, of the peripheral device to be accessed. The second byte is the command byte, which contains the code of the type of operation to be performed. The two general i/o commands are read (52) and write (57). The put (50), status (53), and format (21) commands are used to control disk drives. The third and fourth command bytes contain auxiliary information that is device specific and is used to set operating parameters of the accessed device. The fifth byte of the command frame is the checksum byte used by peripherals to test for transmission errors. The checksum is the sum of data in all bytes sent, plus the carry flag. After the checksum byte is sent the processor pulls the command line high signaling the end of the command frame.

All peripherals on the bus receive the command frame but only that one addressed acknowledges receipt of it. If the checksum computed by the peripheral matches that sent and if the command and auxiliary information are valid, the peripheral responds by sending the acknowledge (ACK) code (41) to the processor on the data input line. If a checksum, command, or auxiliary is detected, the peripheral sends the not acknowledge (NAK) code (4E). If the not acknowledge code is sent, the processor sends the command frame again.

After the peripheral acknowledges the command frame the operation specified begins. If the operation is a data read, the peripheral performs the data access operation and sends the complete code (43) to the processor on the data input line. The peripheral then sends the data bytes requested and finally sends the checksum byte.

If the operation is a data write the processor sends the data bytes and checksum byte to the peripheral on the data output line. The data on the data output line is stable during the falling edge of the processor clock output line. The peripheral receives the data and, after comparing the received and computed checksums, replies on the data input line with an ACK or NAK code. If the NAK is sent to the processor, it sends the data and checksum bytes again. When the peripheral completes the output operation it sends the complete code. Should any of the operations not be successful the peripheral sends the error code (45) and the operation is attempted again.

CHAPTER 15

The TI-99/4 Port

The TI-99/4 home computer is a complete ready-to-use system. The console contains the TI-9900 16-bit CPU, 16K RAM, 26K ROM, video generator, and sound generator. The on-board ROM contains a 14K BASIC plus the video, sound, and monitor routines. There are connectors on the console for two cassette recorders (back), video output (back), two joysticks (left side), solid-state software command modules (top right), and i/o port (right side). The solid-state software command modules contain up to 30K of ROM with educational, entertainment, financial, and other programs. The modules are easily inserted and removed from the connector on top of the console.

This chapter is concerned with the i/o port on the right side of the TI-99/4. This port is used for expansion of RAM or ROM and other peripherals, such as floppy disks, printers, modems, A/D converters, RS-232 interface, speech synthesizers, etc. There is a 44-pin edge connector with numbering as shown in Fig. 15-1. The pinout designations are shown in Fig. 15-2.

TI-99/4 SIGNAL DEFINITIONS

Pin No.	Signal	Definition
37	D0	Data bus—Bidirectional data lines used to transfer data between CPU and i/o devices. D0 is the most significant bit and D7 is the least significant bit. See discussion for further explanation of data transfers.
40	D1	
39	D2	
42	D3	
35	D4	
38	D5	
36	D6	
34	D7	

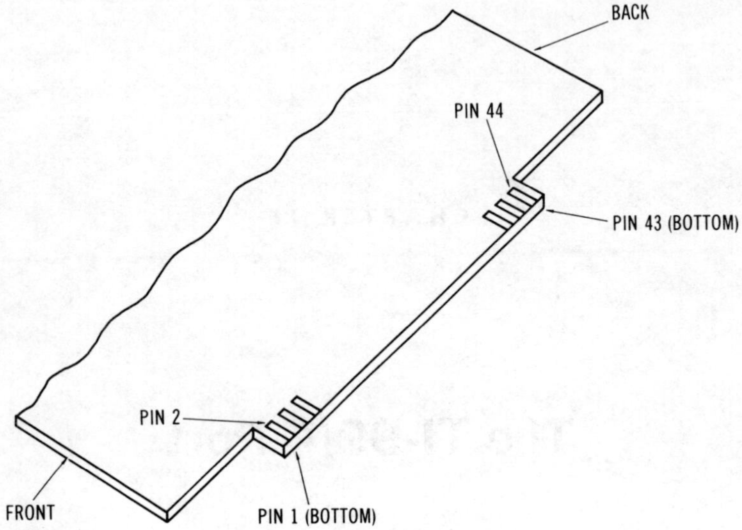

Fig. 15-1. The TI-99/4 pin edge connector.

Pin No.	Signal	Definition
31	A0	Address bus—A0 through A14 are buffered CPU address lines, with A0 being the most significant bit and A14 the least significant bit.
30	A1	
20	A2	
10	A3	
7	A4	
5	A5	
29	A6	
17	A7	
14	A8	
18	A9	
6	A10	
8	A11	
11	A12	
15	A13	
16	A14	
19	A15/CRUOUT	Address 15/CRUOUT—The A15/CRUOUT is used as the least significant address bit for normal data transfers and as the CRUOUT for CRU operations. If $\overline{\text{MEMEN}}$ is true, this line is A15; if $\overline{\text{MEMEN}}$ is false, CRUOUT data appears on this line. See discussion for further explanation of A15 derivation and use.
26	$\overline{\text{WE}}$	Write enable—The $\overline{\text{WE}}$ is the CPU write enable gated with additional clock logic to generate two i/o cycles which are used to transfer 2 bytes of multiplexed data during one CPU transfer. See discussion for i/o read and write cycles.
38	$\overline{\text{MBE}}$	Memory block enable—The $\overline{\text{MBE}}$ indicates a memory access in address block 4000–5FFF. It is the CPU $\overline{\text{MEMEN}}$ gated with A0, A1, and A2.
32	$\overline{\text{MEMEN}}$	Memory enable—The $\overline{\text{MEMEN}}$ indicates a memory access. It is the buffered CPU $\overline{\text{MEMEN}}$.

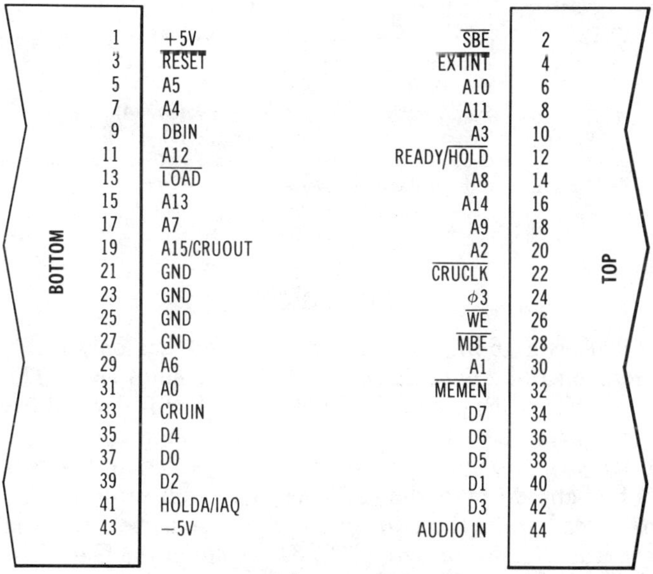

Fig. 15-2. TI-99/4 pinout designations.

Pin No.	Signal	Definition
12	READY/$\overline{\text{HOLD}}$	Ready/hold—The ready/hold is synchronized with CPU clock $\overline{\phi 2}$ and is connected to CPU ready line. If pulled low by a slow i/o device, the CPU will insert wait states until line goes high. Not connected to CPU hold input line, the ready/hold cannot be used for DMA.
9	DBIN	Data bus in—When the DBIN is active, high, the data buffers and the CPU are in the input mode. It is the buffered CPU output line DBIN.
24	$\overline{\phi 3}$	CPU clock—The $\overline{\phi 3}$ signal is an inverted, buffered $\phi 3$ of the CPU clock.
4	$\overline{\text{EXTINT}}$	External interrupt—When active, the $\overline{\text{EXTINT}}$ causes an interrupt request to the CPU $\overline{\text{INTREQ}}$. This is software maskable. The interrupt vector address is 0004.
3	$\overline{\text{RESET}}$	Reset—This signal is the buffered output of a reset circuit which resets the CPU and may be used to initialize i/o devices.
22	$\overline{\text{CRUCLK}}$	CRU clock—The $\overline{\text{CRUCLK}}$ indicates that data is available on the CRUOUT line. It is the inverted, buffered CPU CRUCLK line.
33	CRUIN	CRU data in—Data is input to the CPU's CRU on the CRUIN line.
41	HOLDA/IAQ	Hold acknowledge/instruction acquisition—When $\overline{\text{MEMEN}}$ is active, the HOLD/IAQ indicates that the CPU is acquiring an instruction. When $\overline{\text{MEMEN}}$ is false it indicates that CPU is in a hold state.
2	$\overline{\text{SBE}}$	Speech block enable—The $\overline{\text{SBE}}$ indicates a memory access in speech memory. The output is true on active $\overline{\text{MEMEN}}$ and even addresses between 9000 and 97FE.
44	AUDIO IN	Audio in—The audio in input line for audio signals from the speech module.

Pin No.	Signal	Definition
13	$\overline{\text{LOAD}}$	Load—An active signal causes the CPU to execute a nonmaskable interrupt with address FFFC containing the trap vector. It is the CPU load line.
1	+5V	Supply voltage for speech module (50 mA).
43	−5V	Supply voltage for speech module (50 mA).
21	GND	Ground—Power and signal ground. These four lines are one etch on the edge connector.
23	GND	
25	GND	
27	GND	

DISCUSSION

The 9900 is a 16-bit CPU but the i/o port has only eight lines. Therefore, on 16-bit data transfers the data is multiplexed. The CPU has no A15 (lsb) line, so in the TI-99/4 this line is derived from the clock circuitry. The A15 is high at the beginning of and goes low about halfway through a read or write data access instruction. This selects first an odd byte, then an even byte address on the i/o port. During a read when the odd byte is addressed, the data is latched onto the least significant byte (D15–D8) lines of the CPU. The even byte is then addressed and the data appears at the most significant byte (D7–D0) lines of the CPU. Both bytes are then input to the CPU. During a 16-bit write to the i/o port, the CPU outputs a 16-bit word with A15 selecting the odd byte first, then the even byte. This requires two write signals to the i/o port. The write enable $\overline{\text{WE}}$ on the port is derived from the CPU $\overline{\text{WE}}$ gated with logic from the clock circuitry to develop an active $\overline{\text{WE}}$ while A15 is high and another $\overline{\text{WE}}$ while A15 is low. The clock logic also activates the CPU ready line that inserts wait states while the i/o read or write multiplexing is performed.

The 9900 CPU communications register unit (CRU) is a shift register through which data can be transferred to or from memory or i/o devices serially. Data is input on CRUIN and output on CRUOUT with CRUCLK indicating a valid output bit.

CHAPTER 16

The Digital Group Bus

Although The Digital Group systems are no longer produced, there are several such systems currently in use.

The bus philosophy used in The Digital Group microcomputers, shown in Fig. 16-1, provides for the use of almost any microprocessor chip. A nearly complete set of lines allows operation with an 8080, a Z-80, and a 6800 or 6502 processor board. The data lines are separated into memory input, memory output, peripheral input and peripheral output lines. The address lines are duplicated in memory and peripheral address lines. Interrupt and DMA capabilities are included.

A standard motherboard from The Digital Group, Inc., shown in Fig. 16-2, has connectors for three memory boards, a processor board, a dedicated i/o board, a video terminal board, and three general-

Courtesy The Digital Group, Inc.

Fig. 16-1. A microcomputer by The Digital Group, Inc.

Courtesy The Digital Group, Inc.

Fig. 16-2. A populated motherboard.

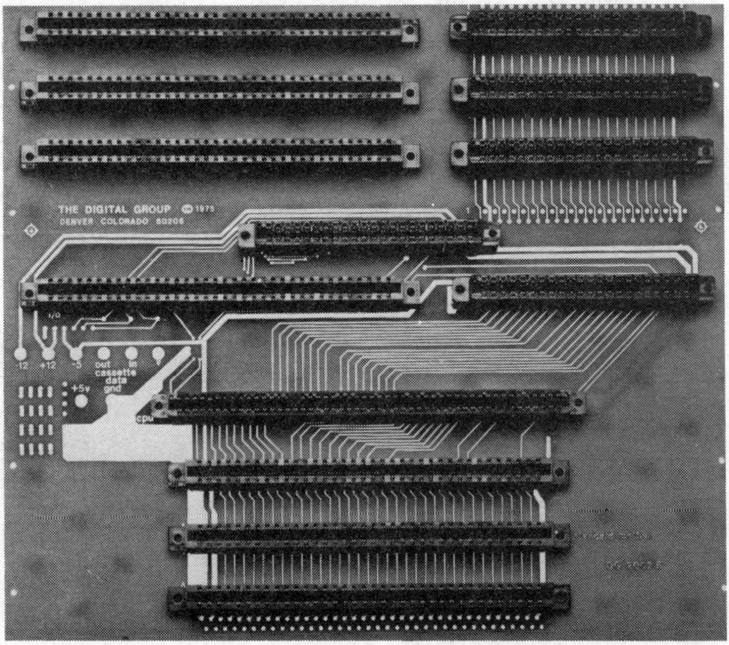

Courtesy The Digital Group, Inc.

Fig. 16-3. The Digital Group's standard motherboard.

purpose i/o boards. This is more clearly seen by the view of the bare motherboard shown in Fig. 16-3.

Although four different type connectors are used on the motherboard, all boards are 12 by 5 inches (30.4 by 12.7 cm). The memory boards plug into 86-pin (43 on each side) double-sided edge-card connectors. The connectors have a 0.156-inch (0.396-cm) spacing. The processor board plugs into a 100-pin, 0.1-inch (0.25-cm), double-sided, edge-card connector. The video board plugs into a 44-pin, 0.156-inch, double-sided, edge-card connector.

Each i/o card has two connectors. The i/o bus signals are brought to a 44-pin, 0.156-inch, double-sided edge-card connector. A 72-pin, 0.156-inch, double-sided edge-card connector is also provided to serve as the connection points for input and output lines. The lines on these connectors are not bused or wired to any other connector on the motherboard. They serve only the ports on the board with which they are associated. Fig. 16-4 shows an open system with the motherboard swung out. Notice the absence of runs to the 72-pin i/o connectors on the lower right side of the motherboard. Positive indexing is not provided for any of the connectors.

External connections to the i/o boards are made through the backplane board mounted in the left side of the back panel shown in Fig. 16-4. Although none are shown, wires join the backplane connectors and the 72-pin i/o connectors on the motherboard. Two 22-pin, double-sided, edge-card connectors are available to the user as well as connectors specified for keyboard, disk, printer, and digi-

Courtesy The Digital Group, Inc.

Fig. 16-4. An open system by The Digital Group, Inc.

tal cassette deck. Additional user i/o connectors can be mounted in holes provided in the back panel. Since system expansion can be accomplished using either the memory or i/o slots, both will be examined.

MEMORY BUS SIGNAL DEFINITIONS

The memory-connector pinout designations are shown in Fig. 16-5.

Pin No.	Signal	Definition
3–10	MI0–MI7	Memory input—The memory input lines carry data from memory to the processor's D0–D7 data lines.
11	$\overline{\text{MRD}}$	Memory read—The memory-read data strobe is an active-low control line that is pulled low by the processor to transfer the data on the memory input lines.
L, K, J, H, F, E, D, C	MO0–MO7	Memory output—The memory output lines carry data from the processor's D0–D7 data lines to memory.
X	$\overline{\text{MWR}}$	Memory write—The memory-write data strobe is an active-low control line that is pulled low by the processor to transfer data on the memory-output lines.
27	$\overline{\text{WAIT}}$	Wait request—The wait line is an active-low input to the processor board. It is used by slow memory to gain more time to set up or retrieve data from the memory input or output lines.
N, P, R, S, T, U, V, W, 12–19	A0–A15	Memory address—The memory address lines are the buffered outputs of the address lines of the processor and carry the number of the desired location to memory.
30	VMA	Valid memory address—The VMA is a processor output that indicates a stable, valid address on the address lines. It is available only in 6800-based systems. In other systems this line is undefined.
AE	$\overline{\text{MEMRQ}}$	Memory request—The memory request is an active-low processor output that is pulled low to indicate a memory operation in Z-80-based systems. It is undefined in configurations using other processors.
29	$\overline{\text{ROM CE}}$	ROM chip enable—The $\overline{\text{ROM CE}}$ is an active-low output used to enable a bootstrap cassette loader ROM placed in hex locations 0000 to 00FF. The bootstrap ROM would normally be used only during system power-up. This line is controlled by an external switch and allows the user to gain those 256 bytes of memory space for RAM use after the loader has been used.
21	$\overline{\text{RFSH}}$	Refresh—The refresh line is an active-low output of the Z-80 during an op-code fetch M1 cycle. It indicates to dynamic RAM refresh circuitry that address lines A0–A6 contain the next refresh row address. The refresh circuitry can then perform the refresh function without slowing the processor.
28	$\overline{\text{RESET}}$	Reset—This line is an active-low input to the processor's reset line. Pulling it low will always reinitialize the system.
AH	$\overline{\text{NMI}}$	Nonmaskable interrupt—The $\overline{\text{NMI}}$ is an active-low input to the processor's $\overline{\text{NMI}}$ line in Z-80, 6800- and 6502-based systems. Pulling it low generates an interrupt request that must be serviced by the processor.

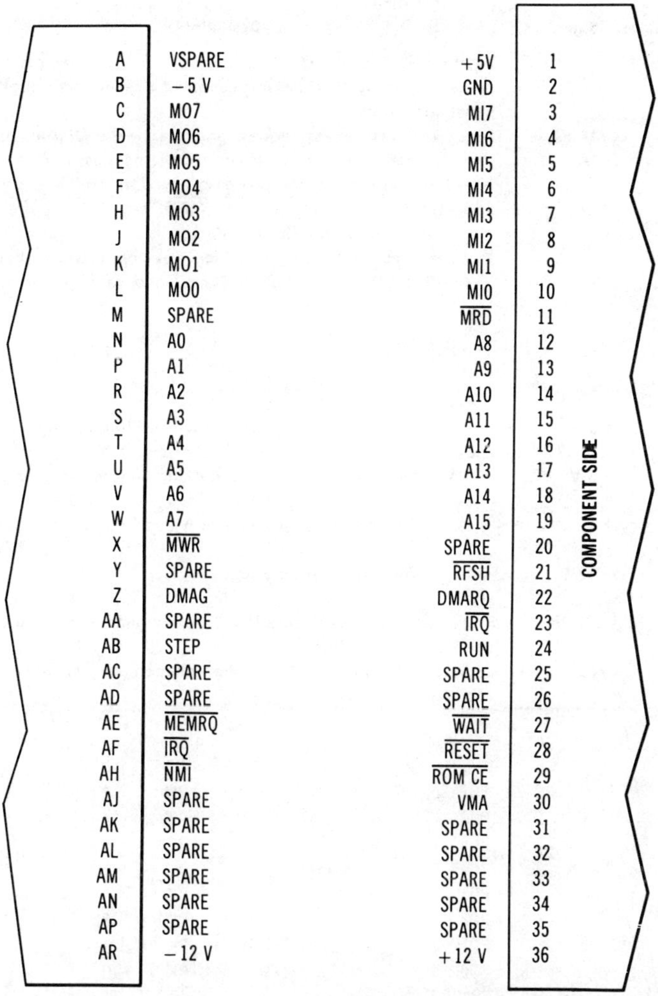

Fig. 16-5. The memory-connector pinout designations.

Pin No.	Signal	Definition
23, AF	$\overline{IRQ}$	Interrupt request—The $\overline{IRQ}$ is an active-low input to the processor's $\overline{IRQ}$ line. Pulling it low causes the processor to check the interrupt status bit and, if not masked, service the request. The "I" status bit is under software control and can be masked to prevent the processor from servicing the request.
24	RUN	Run—The run line is an input that can stop or single step the processor through its $\overline{WAIT}$ line when used with the stop/step input. It must be pulled high to allow normal execution.
AB	STEP	Step—The step/stop line is an input that is used to single-step execution of a program when the processor has been stopped

Pin No.	Signal	Definition
		by the run line. Pulling this line to +5 V will single-step the processor, allowing close inspection of execution through front-panel indicators.
22, Z	DMARQ, DMAG	Direct memory access request and grant—The DMARQ input is used by peripherals to gain control of the buses. When pulled high by a peripheral, the processor stops execution and puts its address, data out, $\overline{MRD}$, $\overline{MWR}$, $\overline{MRQ}$, and $\overline{RFSH}$ buffers in a high-impedance state. The DMAG line is pulled high to indicate that the lines are floating and the DMA operation can begin.
20, 25, 26, 31–35, M, Y, AA, AC, AD, AJ–AP	SPARE	Spare—These lines are not dedicated and can be user defined.
1	+5 V	+5 V—This power-supply line provides filtered, regulated voltages.
B	−5 V	−5 V—This power-supply line provides filtered, regulated voltages.
36	+12 V	+12 V—This power-supply line provides filtered, regulated voltages.
AR	−12 V	−12 V—This power-supply line provides filtered, regulated voltages.
A	VSPARE	Spare voltage—The Vspare is reserved for an additional but yet undefined power line.
2	GND	Ground—The ground line is the common system power-supply and signal-line return.

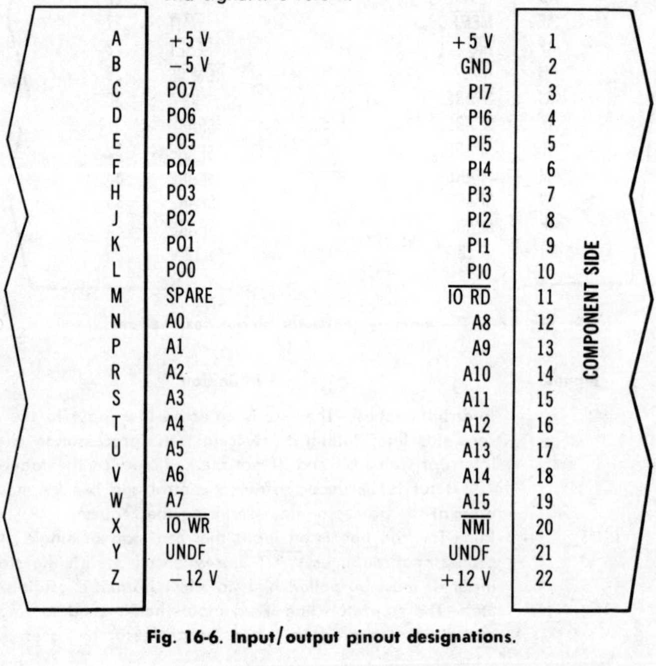

Fig. 16-6. Input/output pinout designations.

I/O SIGNAL DEFINITIONS

The signals available at the i/o connectors are a limited subset of the lines of the processor. They allow use of most parallel and serial interface chips. Memory expansion should occur through the memory connectors. The i/o pinout designations are shown in Fig. 16-6.

Pin No.	Signal	Definition
3–10	PI0–PI7	Peripheral input—The peripheral input lines carry the complement of data from peripherals to the processor's D0–D7 data lines through inverting buffers. The inverters can be replaced to transfer noninverted data.

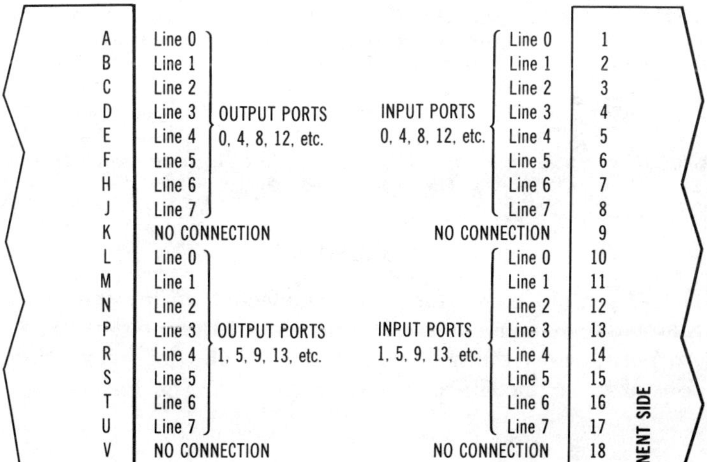

Fig. 16-7. Suggested peripheral pin designations.

Pin No.	Signal	Definition
11	$\overline{\text{IO RD}}$	Input/output read—The i/o read strobe is an active-low control line that is pulled low by the processor to transfer data on the peripheral input lines.
C, D, E, F, H, J, K, L	PO0–PO7	Peripheral output—The peripheral output lines carry data from the processor's D0–D7 data lines to peripherals. These lines are not inverted.
X	$\overline{\text{IO WR}}$	Input/output write—The i/o write strobe is an active-low control line, pulled low by the processor, to transfer data on the peripheral output lines.
12–19, N, P, R, S, T, U, V, W	A0–A15	Peripheral address—The peripheral address lines carry the desired i/o port location from the processor to peripherals. Only A0–A7 are active in Z-80- and 8080-based systems.
20	$\overline{\text{NMI}}$	Nonmaskable interrupt—The $\overline{\text{NMI}}$ is an active-low input to the processor's $\overline{\text{NMI}}$ line on Z-80, 6800- and 6502-based systems. When pulled low, the processor must service the request.
21, Y	UNDF	CPU—I/o undefined—These lines are reserved for processor-i/o signals but are not yet defined.
M	SPARE	Spare—The spare is an unspecified undefined line available to the user.
1, A, B, 22, Z	+5 V, −5 V, +12 V, −12 V	+5 V, −5 V, +12 V, −12 V—These power-supply lines provide filtered, regulated voltages.

DISCUSSION

The 72-pin i/o connector is not connected to any internal lines but provides convenient connection from external devices to onboard peripheral circuitry. Suggested port- and line-number designations of these unconnected pins are given in Fig. 16-7.

CHAPTER 17

The LSI-11 and H-11 Bus

Heath Company's H-11 computer, shown in Fig. 17-1, is LSI-11* based and is plug compatible with DEC's* LSI-11 bus. It uses the DEC LSI-11 processor and executes the PDP-11/03 instruction set. Heath supports the H-11 with a variety of peripheral boards, and the system can accept other LSI-11 compatible boards.

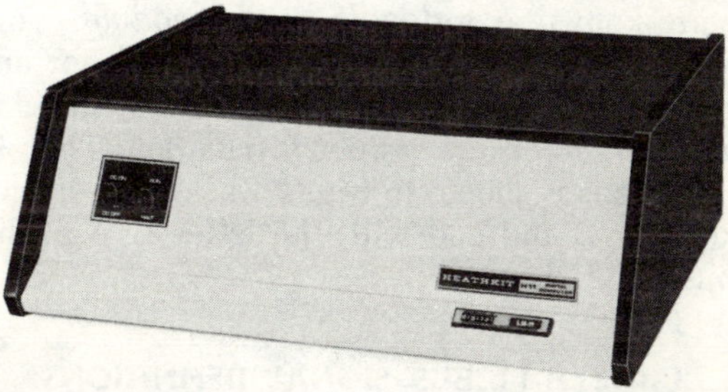

Courtesy Heath Co.

Fig. 17-1. The Heath H-11 computer.

The LSI-11 bus consists of four inline 36-pin edge connectors shown in Fig. 17-2. The connectors are designated A, B, C, and D. The pins on each connector are designated A, B, C, . . . , V, excluding G, I, O, and Q. Pins on the component side of the board are identified by a 1; those on the solder side are identified by a 2. Each pin is identified by three characters: the connector letter, the pin letter,

*LSI-11 and DEC are trademarks of Digital Equipment Corporation.

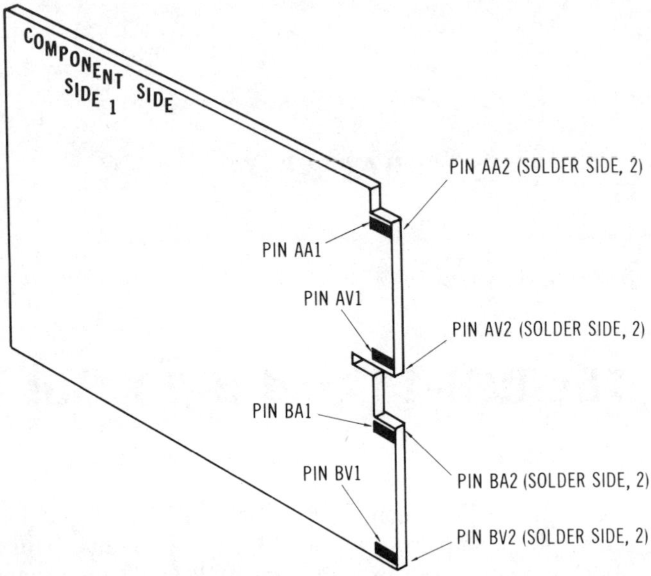

Fig. 17-2. An LSI-11 board.

and the side number. Pin BD1 is the fourth pin on the second connector on the component side of the board. Full-size boards, like that in Fig. 17-2, are known as a *double height module.* I/o boards are only half size and have only connectors A and B. They are known as *double height modules* or *standard modules.* Boards are notched between connectors to ensure proper installation. The pinout of the LSI-11 and H-11 bus is shown in Fig. 17-3. All bus signal names begin with the letter B and end with the letter L, e.g., the BSYNCL line is the Bus SYNC Line.

LSI-11/H-11 BUS SIGNAL DEFINITIONS

Pin No.	Signal	Definition
AU2 AV2 BA2–BV2	BDAL0L– BDAL15L	Data/address lines 0–15—The data/address lines are multiplexed to carry both address and program data. The processor, or other bus master, indicates that the data/address lines hold an address by pulling the sync line high. The processor indicates a read or write operation by pulling the DIN or DOUT line high. Data transfers on the bus are synchronous so a peripheral (memories or other slaves) must signal the processor (master) that it has placed data on or taken data from the bus by pulling the BRPLY line high.
AJ2	BSYNCL	Sync—The sync line is an active-high processor output which indicates that a valid memory address has been placed on the data/address lines.

COMPONENT SIDE — SOLDER SIDE

CONNECTOR A

Pin	Signal	Signal	Pin
AA1	BSPARE1	+5	AA2
AB1	BSPARE2	−12	AB2
AC1	BAD16	GND	AC2
AD1	BAD17	+12	AD2
AE1	SSPARE1	BDOUT	AE2
AF1	SSPARE2	BRPLY	AF2
AH1	SSPARE3	BDIN	AH2
AJ1	GND	BSYNC	AJ2
AK1	MSPAREA	BWTBT	AK2
AL1	MSPAREA	BIRQ	AL2
AM1	GND	BIAKI	AM2
AN1	BDMR	BIAKO	AN2
AP1	BHALT	BBS7	AP2
AR1	BREF	BDMGI	AR2
AS1	PSPARE3	BDMGO	AS2
AT1	GND	BINIT	AT2
AU1	PSPARE1	BDAL0	AU2
AV1	+5B	BDAL1	AV2

CONNECTOR B

Pin	Signal	Signal	Pin
BA1	BDCOK	+5	BA2
BB1	DPOK	−12	BB2
BC1	SSPARE4	GND	BC2
BD1	SSPARE5	+12	BD2
BE1	SSPARE6	BDAL2	BE2
BF1	SSPARE7	BDAL3	BF2
BH1	SSPARE8	BDAL4	BH2
BJ1	GND	BDAL5	BJ2
BK1	MSPAREB	BDAL6	BK2
BL1	MSPAREB	BDAL7	BL2
BM1	GND	BDAL8	BM2
BN1	BSACK	BDAL9	BN2
BP1	BSPARE6	BDAL10	BP2
BR1	BEVNT	BDAL11	BR2
BS1	PSPARE4	BDAL12	BS2
BT1	GND	BDAL13	BT2
BU1	PSPARE2	BDAL14	BU2
BV1	+5	BDAL15	BV2

Fig. 17-3. The H-11 and LSI-11 pinout designations.

Pin No.	Signal	Definition
AH2	BDINL	Data input line—The DIN line is an active-high output used by the processor or other bus master during sync to indicate that a memory read or an input operation is about to occur. The processor requires a reply, through RPLY, to end an input operation. The processor also uses the DIN line to indicate that an interrupt is being processed by pulling the DIN line high while the sync line is low.
AK2	BWTBTL	Write/byte—The write/byte is a dual-function, active-high processor output. When pulled high during sync the write/byte indicates that a memory write, an output operation, is about to occur. If the write/byte line is pulled high during data out, the processor is indicating a byte address.

Pin No.	Signal	Definition
AE2	BDOUTL	Data output—The DOUT line is an active-high processor output that is pulled high to indicate that a memory write, an output operation, is about to occur. The processor demands a reply, through RPLY, from the peripheral to end the output operation.
AF2	BRPLYL	Reply—The RPLY line is an active-high processor input used by peripherals, memories, or other slaves, to signal the processor that data has been placed on the data/address lines during DIN, or that data has been removed from the data/address lines during DOUT. The processor must receive the RPLY signal to drop the DIN or DOUT and end the data transfer operat:on.
AT2	BINITL	Initialize—The INIT line is an active-high processor output that is used to initialize, or reset, devices on the bus after a power-up operation.
AN1	BDMRL	Direct memory access request—The DMR is an active-high processor input used by peripherals to request control of the bus for DMA operations. The processor, at the end of the current bus cycle, recognizes the request with the DMGO signal.
AS2 AR2	BDMGOL BDMGIL	Direct memory access grant output and direct memory access grant input—The DMSO and DMSI lines are used to form a daisy chain priority DMA system. The processor's DMGO is connected to the highest priority board's DMGI. The highest priority board's DMGO is connected to the next lower priority board's DMGI line, etc. When the processor grants a DMA request it pulls the DMGO line high. The highest priority board receives the signal, and if it requested control of the bus, it does not pass that signal to the next board. If the highest priority board did not request the bus, it passes the DMGO on. The DMGO is passed through the boards until it is received by the requesting board.
BN1	BSACKL	Bus acknowledge—The BSACK line is an active-high processor input that is used by the peripheral requesting control of the bus, via DMR, that it has received DMGO, via the daisy chain, and has taken control of the bus. The DMA ends when the peripheral pulls the BSACK low and the processor regains control of the bus.
AP1	BHALTL	Processor halt—The HALT line is an active-high processor output used to indicate that the processor has suspended execution and is waiting for a DMA request. While in the halt mode the processor will not respond to interrupt requests but will continue to refresh memory.
AL2	BIRQL	Interrupt request—The IRQ line is an active-high processor input used by peripherals to request service only if its interrupt enable and interrupt request flip-flops are set. The processor will acknowledge an interrupt if bit 7 of its PS word is zero by pulling the DIN and IAKO lines high.
AM2 AN2	BIAKIL BIAKOL	Interrupt acknowledge input and interrupt acknowledge output—The IAKO and IAKI lines are used to form a daisy-chain interrupt priority system. The processor responds to an interrupt request, via IRQ, by pulling its IAKO line high. This line is connected to the highest priority board's IAKI line. The highest priority board's IAKO line is connected to the next highest board's IAKI line, etc., to form a daisy chain. A higher

Pin No.	Signal	Definition
		priority board not requesting service passes the IAK signal to the next lower priority board. When the board that requested service receives the IAK signal it places the proper vector on the DA lines then pulls the RPLY high.
BR1	DEVNTL	External event interrupt request—The EVNT is an active-high processor input that is used by devices to request interrupt service. The processor responds to this signal by fetching a vector from location 1141 and beginning execution of the service routine at the vector address. The EVNT is typically used as a real-time clock input.
AR1	BREFL	Memory refresh—The REF line is an active-high processor or peripheral output that indicates a refresh cycle for dynamic memories on the bus.
AP2	BBS7L	Bank 7 select—The BS7 is an active-high processor output used to indicate that the address on the bus is in the upper 4K bank. The processor follows the BS7 signal with the SYNC signal.
AC1 AD1	BAD 16, 17	Extended address bits 16 and 17—The AD 16 and AD 17 lines are active-high processor outputs that may be used to extend the addressing range.
BA1	BDCOKH	DC power OK—The DCOK line is an active-high processor input that is generated by the power supply and indicates that power is available to support reliable system operation.
BB1	BPOKH	Power OK—The POK signal is an active-high processor input that is generated by the power supply and indicates that primary power is normal. If this line is pulled low during normal operation, a power fail trap sequence is initiated.
AE1, AF1, AH1, BC1, BD1, BE1, BF1, BH1	SSPARE 1– SSPARE 8	Special spare 1–8—The SSPARE lines are not interconnected and are reserved for customer use.
AA1, AB1, BP1	BSPARE 1 BSPARE 2 BSPARE 6	Bus spare 1, 2, 6—The SPARE lines are not assigned but are reserved for manufacturer's use.
AK1, AL1, BK1, BL1	MSPAREA MSPAREB	Maintenance spare 1, 2—The MSPAREs are normally interconnected.
AU1 BU1 BS1	PSPARE 1 PSPARE 2 PSPARE 4	Spare 1, 2, 4—These SPARE lines are not assigned but are not recommended for customer use.
BV1, AA2, BA2	+5	+5—The +5 line is the system +5-volt dc supply line.
AD2, BD2	+12	+12—The +12 line is the system +12-volt dc supply.
AB2, BB2	−12	−12—The −12 line is the system −12-volt dc supply.
AV1	+5B	+5-V battery power—The +5B line supplies battery backup +5 volts dc.
AS1	+12B	+12-V battery power—The +12B line supplies battery backup +12 volts dc.
AJ1, AM1, AT2, BJ1, BM1, BT1, AC2, BC2	GND	Ground—The system supply and signal ground lines.

CHAPTER 18

The Multibus

The Multibus was developed by Intel Corporation for their SBC series of boards that includes CPUs, memory, i/o, communications, high-speed math, DMA control, and others. Fig. 18-1 shows an SBC system with chassis, power supply, backplane with card cage, and CPU board. Figs. 18-2 and 18-3 show an SBC 80/20 CPU board and a combination memory and i/o expansion board, respectively. Several companies produce a wide variety of boards which will operate on the Multibus.

The SBC series is a general-purpose, inexpensive board set originally designed for oem's so that they would not have to develop their own hardware for every new product in which they used an Intel

Courtesy Intel Corp.

Fig. 18-1. The SBC-based 80/20 system.

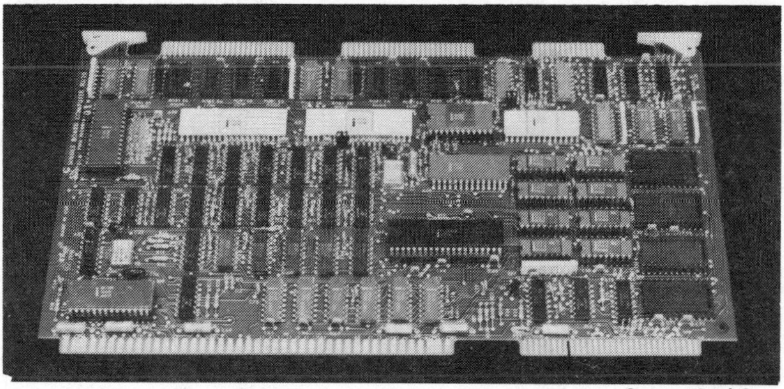

Courtesy Intel Corp.

Fig. 18-2. SBC 80/20 single-board computer.

processor. The SBC system allows implementation of a tremendous variety of inputs and outputs with a small number of boards. Multimaster operation having multiplexed CPUs, DMA controllers, and disk controllers, controlling the bus is easily realized.

A typical board for the Multibus is shown in Fig. 18-3. It is 12 inches (30.4 cm) long, 6.75 inches (17.1 cm) high, and 0.062 inch (0.05 cm) thick with two edge connectors along the bottom as shown. (The P1 Multibus is on the left in the photo and the P2 Auxiliary Connector is on the right.) Depending on the board type,

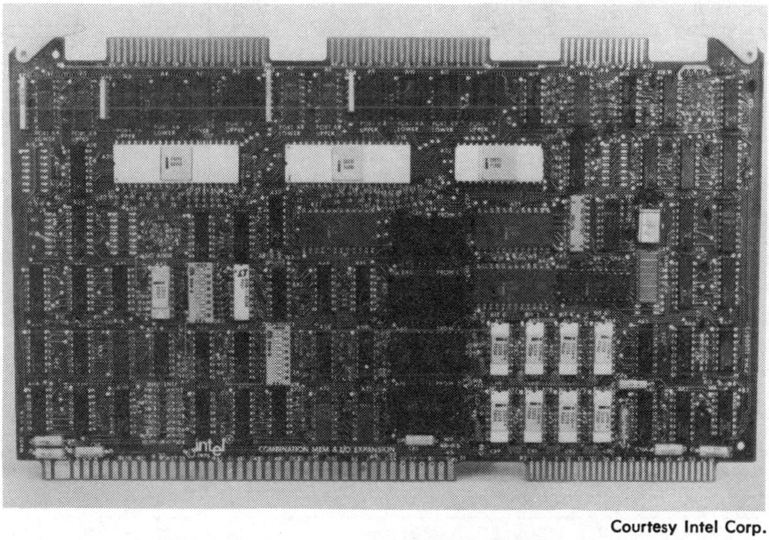

Courtesy Intel Corp.

Fig. 18-3. SBC 517 combination i/o expansion board.

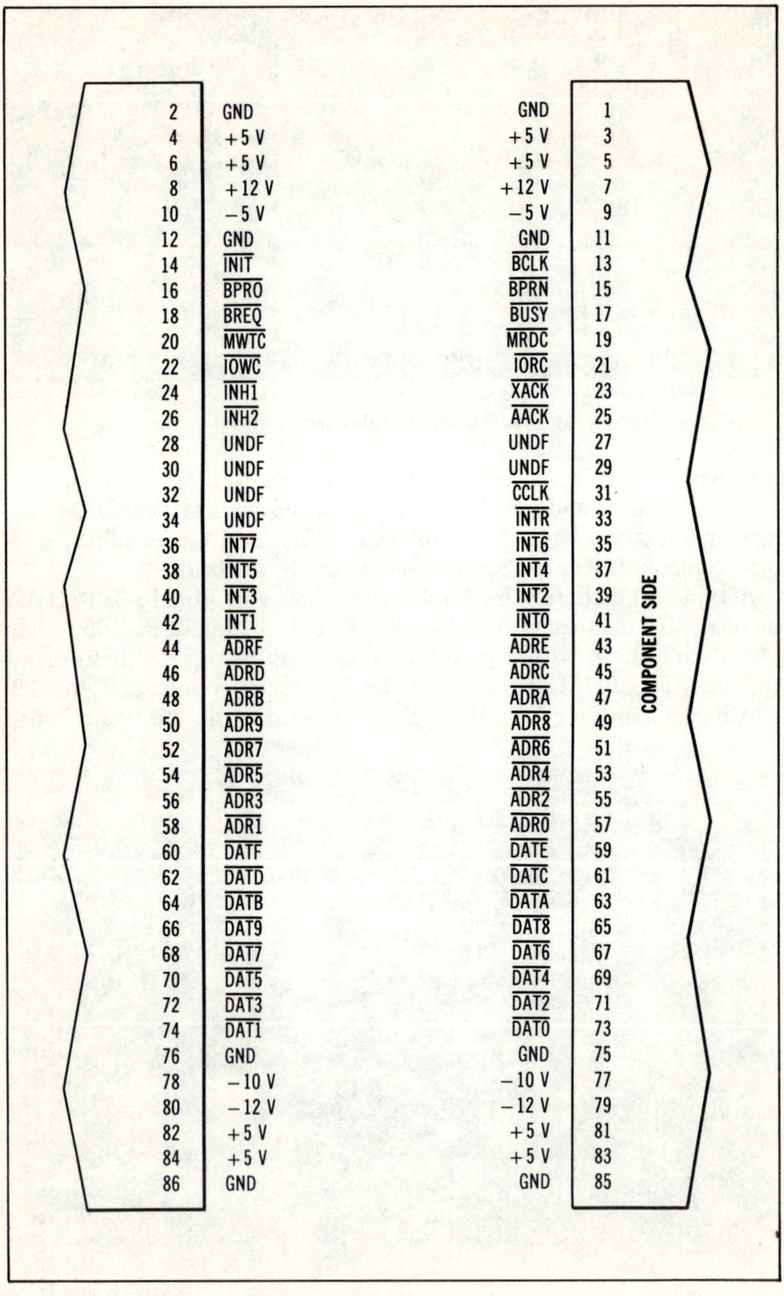

Fig. 18-4. SBC pinout designations.

there may be up to three edge connectors along the top of the board. The 86-pin P1 (Multibus) edge-connector pins are numbered from left to right as the board is viewed in Fig. 18-3. The odd pin numbers are on the component side and the even pin numbers are on the opposite side. The pinout designations are listed in Fig. 18-4.

SBC SIGNAL DEFINITIONS

Pin No.	Signal	Definitions
67–74	$\overline{DAT0}$–$\overline{DAT7}$	Data bus—Active low, bidirectional.
59–66	$\overline{DAT8}$–$\overline{DATF}$	Data bus—For upper 8 bits in 16-bit systems.
19	$\overline{MRDC}$	Memory read command—A low indicates that a certain memory location is on the address bus and its contents are to be put on data bus.
20	$\overline{MWTC}$	Memory write command—A low indicates that a certain memory location has been addressed and data to be written into it is on the data bus.
21	$\overline{IORC}$	I/o read command—A low indicates that an input port has been addressed and the data at that port is to be placed on the data bus.
22	$\overline{IOWC}$	I/o write command—A low indicates that an output port has been addressed and that the data bus contents are to be outputted.
43–58	$\overline{ADR0}$–$\overline{ADRF}$	Address bus—Active-low address lines.
14	$\overline{INIT}$	Initialize—A low resets the entire system.
15	$\overline{BPRN}$	Bus priority in—A low indicates that no higher priority master module is requesting use of the system bus. It is connected to signal ground in a single master system; synchronized with $\overline{BCLK}$ in a multimaster system. It is not bused on the motherboard.
16	$\overline{BPRO}$	Bus priority out—Connected to $\overline{BPRN}$ of master module with next lower bus priority. A low indicates to the next lower master module that a higher priority master is in control of the buses. It is not bused on motherboard.
17	$\overline{BUSY}$	Bus busy—A low indicates that the bus is currently in use and prevents other master modules from gaining control.
18	$\overline{BREQ}$	Bus request—A low indicates that a particular master module requires use of the bus. It is synchronized with $\overline{BCLK}$.
23	$\overline{XACK}$	Transfer acknowledge—A low from a memory location or i/o port indicates that the specified read/write operation is complete.
25	$\overline{AACK}$	Advanced acknowledge—A low allows memory to complete access without having the CPU wait.
33	$\overline{INTR}$	Direct interrupt—A low from an external device generates an interrupt request.
35–41	$\overline{INT0}$–$\overline{INT7}$	Parallel interrupt requests—A low from an external device generates an interrupt request.
13	$\overline{BCLK}$	Bus clock—Used to synchronize bus control circuits on all master modules. It may be slowed, stopped, or single stepped.
31	$\overline{CCLK}$	Constant clock—Provides a clock signal of constant frequency for use by optional memory and i/o boards.
1, 2, 11, 12, 75, 76, 85, 86	GND	Signal ground.

Pin No.	Signal	Definitions
3–6, 81–84	+5 V	+5-Vdc regulated voltage.
7, 8	+12 V	+12-Vdc regulated voltage.
9, 10	−5 V	−5-Vdc regulated voltage.
79, 80	−12 V	−12-Vdc regulated voltage.
24	$\overline{\text{INH1}}$	Inhibit 1—Disables RAM memory devices. Allows ROM to override RAM in same address.
26	$\overline{\text{INH2}}$	Inhibit 2—Disables ROM memory devices. Allows auxiliary ROM to override ROM in same address.
27–30, 32, 34	UNDF	Undefined—These lines are undefined on the bus.
77, 78	−10 V	−10 Vdc (for MDS 800 compatibility).

CHAPTER 19

The Exorcisor Bus

Motorola introduced the Exorcisor as a system development and debugging aid for their 6800 microprocessor and its family of support chips. It is widely used in industry and has grown in popularity among hobbyists. A wide variety of system configurations is available, ranging from single-board computers, like the 68MM 01A shown in Fig. 19-1, to expanded disk-based systems. The tty-based EVKI and hex keypad and display-based EVKII (shown unassem-

Courtesy Motorola, Inc.

Fig. 19-1. An Exorcisor-compatible micromodule system.

bled in Fig. 19-2) have been popular hobbyist systems. Over 20 companies produce Exorcisor compatible boards, of which most are data conversion boards.

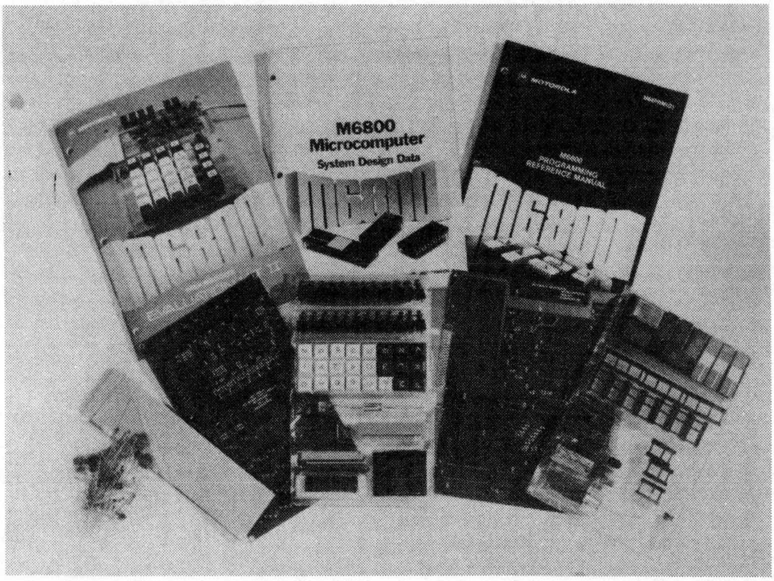

Courtesy Motorola, Inc.

Fig. 19-2. Motorola's 6800 Evaluation Kit II.

Exorcisor boards are 5¾ by 9¾ inches (14.6 by 24.7 cm). They use an 86-pin (43 on each side) edge connector as shown in Fig. 19-1. No positive indexing is provided on all system boards.

EXORCISOR BUS SIGNAL DEFINITIONS

The Exorcisor bus pinout designations are shown in Fig. 19-3.

Pin No.	Signal	Definition
29–32, $\overline{\text{H}}$, $\overline{\text{J}}$, $\overline{\text{K}}$, $\overline{\text{L}}$	$\overline{\text{D0}}$–$\overline{\text{D7}}$	Data bus—The bidirectional data bus lines over which data passes between the processor and the peripherals.
6	R/W	Read/write—The read/write lines are an output from the processor that informs peripherals of the direction of data transfer. It is high during a read operation and low during a write operation.
33–40, $\overline{\text{M}}$, $\overline{\text{N}}$, $\overline{\text{P}}$, $\overline{\text{R}}$–$\overline{\text{V}}$	A0–A15	Address bus—The address bus lines are outputs from the processor that carry the desired address to the peripherals.
F	VMA	Valid memory address—The valid memory address line is a processor output that is high when the address bus lines contain a valid address (normally during the clock $\phi 1$). In an Exorcisor

1	+5V	+5V	A
2	+5V	+5V	B
3	+5V	+5V	C
4	GO/HLT	$\overline{\text{IRQ}}$	D
5	$\overline{\text{RESET}}$	$\overline{\text{NMI}}$	E
6	R/W	VMA	F
7	$\phi1$	GND	H
8	GND	$\phi2$	J
9	GND	GND	K
10	VUA	MEM CLK	L
11	−12V	−12V	M
12	$\overline{\text{REF REQ}}$	TSC	N
13	REF GNT	BA	P
14	UNDF	MEM RDY	R
15	UNDF	REF CLK	S
16	+12V	+12V	T
17	UNDF	UNDF	U
18	UNDF	UNDF	V
19	UNDF	UNDF	W
20	UNDF	UNDF	X
21	UNDF	UNDF	Y
22	UNDF	UNDF	Z
23	UNDF	UNDF	$\overline{A}$
24	UNDF	UNDF	$\overline{B}$
25	UNDF	UNDF	$\overline{C}$
26	UNDF	UNDF	$\overline{D}$
27	UNDF	UNDF	$\overline{E}$
28	UNDF	UNDF	$\overline{F}$
29	$\overline{\text{D1}}$	$\overline{\text{D3}}$	$\overline{H}$
30	$\overline{\text{D5}}$	$\overline{\text{D7}}$	$\overline{J}$
31	$\overline{\text{D0}}$	$\overline{\text{D2}}$	$\overline{K}$
32	$\overline{\text{D4}}$	$\overline{\text{D6}}$	$\overline{L}$
33	A15	A14	$\overline{M}$
34	A12	A13	$\overline{N}$
35	A11	A10	$\overline{P}$
36	A8	A9	$\overline{R}$
37	A7	A6	$\overline{S}$
38	A4	A5	$\overline{T}$
39	A3	A2	$\overline{U}$
40	A0	A1	$\overline{V}$
41	GND	GND	$\overline{W}$
42	GND	GND	$\overline{X}$
43	GND	GND	$\overline{Y}$

Fig. 19-3. The Exorcisor bus pinout designations.

Pin No.	Signal	Definition
		system the VMA feeds only the executive DEBUG module, which generates the VUA signal.
10	VUA	Valid user address—The valid user address line is a DEBUG module output that is high when the address bus contains a valid address for any location other than one in the EXBUG monitor. It replaces the VMA for peripherals.
5	$\overline{\text{RESET}}$	Reset—The reset line is the output of a one-shot used to reset the processor and peripherals. An input cannot be applied to this line without circuit modification.
E	$\overline{\text{NMI}}$	Nonmaskable interrupt—The nonmaskable interrupt input feeds the processor's $\overline{\text{NMI}}$. When pulled low by a peripheral, the processor finishes the current instruction, stores the internal registers on the stack, loads the $\overline{\text{NMI}}$ vector into the program counter, and begins execution of the $\overline{\text{NMI}}$ service routine.
D	$\overline{\text{IRQ}}$	Interrupt request—The interrupt request input feeds the IRQ of the processor. When pulled low by a peripheral, the processor finishes the current instruction, checks the "I" flag in the CCR and, if it is reset, stores the internal registers on the stack. It then loads the $\overline{\text{IRQ}}$ vector into the program counter and begins execution of the $\overline{\text{IRQ}}$ service routine.
4	GO/HLT	Go/halt—The go/halt line is an input to the halt line of the processor. When pulled low, it causes the processor to suspend all operations, pull the BA high, pull the VMA low, and put all other lines in a high-impedance state.
P	BA	Bus available—The bus available line is an output that indicates whether the processor is running or has been stopped. It is low when the processor is executing a program. It is pulled high when the processor is stopped by the halt input or the execution of a wait instruction. By watching the BA line, an external device can verify that the processor has stopped and its lines are floating, thus allowing DMA.
N	TSC	Three-state control—The three-state control line is an input to the processor that, when pulled high, causes the address bus and R/W line to float and pulls the VMA and BA low. An external device can then use the address bus in DMA operations. The system clock must also be stopped with $\phi1$ high and $\phi2$ low. Unfortunately, the clock can be stopped for only a short period since the processor is a dynamic chip.
7, J	$\phi1$, $\phi2$	$\phi1$, $\phi2$—The two-phase, nonoverlapping system clock which synchronizes all operations.
R	MEM RDY	Memory ready—The memory ready line is an input to the system clock circuitry that allows the processor to work with slow memories by stretching the $\phi2$ signal. (Stretching a clock cycle is momentarily stopping the clock while both $\phi1$ and $\phi2$ signals are held steady.) When pulled low by address decoding logic associated with slow memory, the $\phi2$ clock signal is lengthened. This gives the memory more time to present data to or take data from the data bus. The $\phi2$ signal is stretched since all data transfers occur during its occurrence.
L	MEM CLK	Memory clock—The memory clock line is a $\phi2$ output of the system clock circuitry. It leads the $\phi2$ signal by about 10 ns. It is used by dynamic memory circuits to synchronize refresh

Pin No.	Signal	Definition
		operations with the processor. It is not affected by clock stretching signals.
S	$\overline{\text{REF CLK}}$	Refresh clock—The refresh clock is the output of an astable multivibrator in the dynamic memory refresh circuitry. It is used to generate the $\overline{\text{REF REQ}}$ signal and to control the refresh counters.
12	$\overline{\text{REF REQ}}$	Refresh request—The refresh request line is an input to the system clock that is driven by the refresh circuitry on a dynamic memory board. The $\overline{\text{REF REQ}}$ line is pulled low periodically to inform the processor of the need to refresh dynamic RAM. The processor responds to this signal with the REF GNT, which sets the $\overline{\text{REF REQ}}$ line high, clearing it.
13	REF GNT	Refresh grant—The refresh grant line is an output of the system clock generator that acknowledges an input on the $\overline{\text{REF REQ}}$ line. When the $\overline{\text{REF REQ}}$ is received, the REF GNT is pulled high at the beginning of the next clock $\phi1$. This signal clears the $\overline{\text{REF REQ}}$ and indicates that the cycle beginning can be used as a refresh cycle. The $\phi1$ signal is stretched during the refresh and the processor continues program execution during the following $\phi2$ cycle. Refreshing dynamic RAM in this manner is known as ***cycle stealing***.
1, 2, 3, A, B, C	+5 V	+5 V—The 5-Vdc line is a regulated power input that should be able to supply 15 amperes.
16, T	+12 V	+12 V—The +12-Vdc line is a regulated power input that should be able to supply 2.5 amperes.
11, M	−12 V	−12 V—The −12-Vdc line is a regulated power input that should be able to supply 1.0 ampere.
8, 9, 41-43, H, K, $\overline{\text{W}}$-$\overline{\text{Y}}$	GND	Ground—The ground line is the common return for all system supplies and signals.
14, 15, 17–28, U–Z, $\overline{\text{A}}$–$\overline{\text{F}}$	UNDF	Undefined—These lines are undefined on the bus and may be dedicated to special functions by the user.

CHAPTER 20

The Versabus

Motorola developed the Versabus to accommodate a wide variety of processors including the 8-bit 6800, the 16-bit 68000, and the to-be-announced 32-bit machine. The Versabus has 260 pins and requires rather large, 14.5- by 9.14-inch (36.8- by 23.2-cm), full-size boards, as shown in Fig. 20-1. The bus is divided between two edge connectors: P1 and P2. Connector P1 includes those signal lines needed for 8- and 16-bit processor operation, while P2 includes the

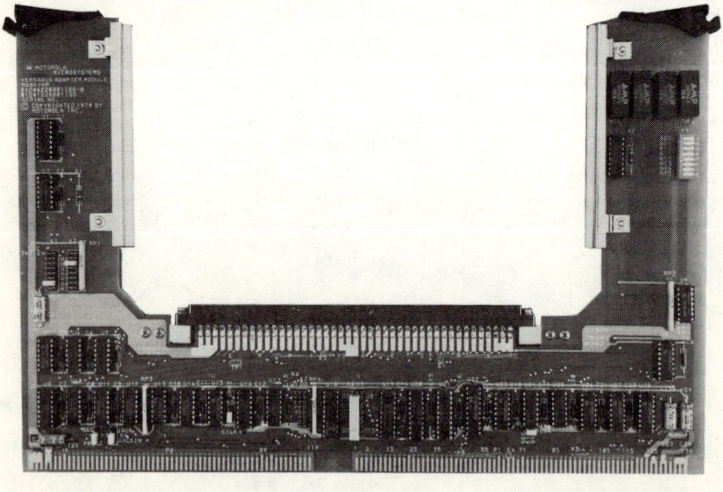

Courtesy Motorola, Inc.

Fig. 20-1. Versabus board that allows operation of smaller Exorcisor boards.

additional data and address lines needed for 32-bit processor operation as well as i/o lines. An 8- or 16-bit processor can be implemented using only connector P1 on an 8- by 9.25-inch (20.3- by 23.5-cm) board. The Versabus pinout is given in Figs. 20-2 and 20-3.

VERSABUS SIGNAL DEFINITIONS

Connector and Pin No.	Signal	Definition
P1 5–20	$\overline{\mathrm{D00}}$–$\overline{\mathrm{D15}}$	Data bus 0–15—The data bus 0–15 lines are active-low bidirectional lines that carry data and instructions between the processor and all peripherals.
P2 105–120	$\overline{\mathrm{D16}}$–$\overline{\mathrm{D31}}$	Data bus 16–31—The data bus 16–31 lines to be used in 32-bit systems.
P1 25–26 P2 99–100	$\overline{\mathrm{DS0}}$–$\overline{\mathrm{DS3}}$	Data strobe 0–3—The data strobe lines are active-low lines used by the bus master to indicate on which data bus lines data will be transferred. Data will be transferred on lines $\overline{\mathrm{D00}}$–$\overline{\mathrm{D07}}$ when DS0 is low; on $\overline{\mathrm{D08}}$–$\overline{\mathrm{D15}}$ when $\overline{\mathrm{DS1}}$ is low; on $\overline{\mathrm{D16}}$–$\overline{\mathrm{D23}}$ when $\overline{\mathrm{DS2}}$ is low; and on $\overline{\mathrm{D24}}$–$\overline{\mathrm{D31}}$ when $\overline{\mathrm{DS3}}$ is low. By using a combination of data strobe lines an 8-, 16-, or 32-bit data transfer can be performed.
P1 21–22 P2 103–104	DPARITY0– DPARITY3	Data parity 0–6—The data parity lines are active-low lines used to indicate odd parity on the 4 data bytes, $\overline{\mathrm{D00}}$–$\overline{\mathrm{D31}}$. Data parity line 0 is used with $\overline{\mathrm{D00}}$–$\overline{\mathrm{D07}}$; data parity line 1 is used with $\overline{\mathrm{D08}}$–$\overline{\mathrm{D15}}$; data parity line 2 is used with $\overline{\mathrm{D16}}$–$\overline{\mathrm{D23}}$; data parity line 3 is used with $\overline{\mathrm{D24}}$–$\overline{\mathrm{D31}}$.
P1 36–58	$\overline{\mathrm{A01}}$–$\overline{\mathrm{A23}}$	Address bus 01–23—The address bus 01–23 lines are active-low bus master outputs used by the master to indicate the location needed in the current operation.
P2 89–96	$\overline{\mathrm{A24}}$–$\overline{\mathrm{A31}}$	Address bus 24–31—The address bus 24–31 lines are active-low address lines to be used in 32-bit systems.
P1 33 P2 88	$\overline{\mathrm{APARITY0}}$ $\overline{\mathrm{APARITY1}}$	Address parity 0 and 1—The address parity 0 and 1 lines are used with $\overline{\mathrm{A01}}$–$\overline{\mathrm{A23}}$ and $\overline{\mathrm{A24}}$–$\overline{\mathrm{A31}}$, respectively, to indicate odd parity.
P1 60	$\overline{\mathrm{EXTDADR}}$	Extended address—The extended address is an active-low bus master output that is used to indicate that the address requested is beyond the lowest 16 megabytes.
P1 30	$\overline{\mathrm{AS}}$	Address strobe—The address strobe is an active-low bus master output that is pulled low when the address lines have settled and hold a valid address.
P1 34	$\overline{\mathrm{WRITE}}$	Write—The write line is an active-low bus master output that is used to indicate a read or write operation. The write is pulled high for a read operation and pulled low for a write operation.
P1 29	$\overline{\mathrm{DTACK}}$	Data transfer acknowledge—The $\overline{\mathrm{DTACK}}$ line is an active-low bus master input that is produced by the peripheral involved in the current processor operation. It is pulled low by a peripheral during a write cycle to indicate that data has been latched from the data bus and it is pulled low during a read cycle to indicate that data has been placed on the data bus. The bus

Pin	Component Side	Solder Side	Pin
1	+5V	+5V	2
3	GND	GND	4
5	D00	D01	6
7	D02	D03	8
9	D04	D05	10
11	D06	D07	12
13	D08	D09	14
15	D10	D11	16
17	D12	D13	18
19	D14	D15	20
21	DPARITY0	DPARITY1	22
23	GND	GND	24
25	DS0	DS1	26
27	GND	GND	28
29	DTACK	AS	30
31	GND	GND	32
33	APARITY0	WRITE	34
35	RESERVED	A01	36
37	A02	A03	38
39	A04	A05	40
41	A06	A07	42
43	A08	A09	44
45	A10	A11	46
47	A12	A13	48
49	A14	A15	50
51	A16	A17	52
53	A18	A19	54
55	A20	A21	56
57	A22	A23	58
59	IOEN	EXTDADR	60
61	GND	GND	62
63	SECMAP	ROMDIS	64
65	PDOWN	RAMDIS	66
67	GND	GND	68
69	ACCLK	SYSCLK	70
71	GND	GND	72
73	SECRESET	SYSRESET	74
75	SYSPAUSE	SYSPACK	76
77	SECIEN	ACFAIL	78
79	TEST	SYSFAIL	80
81	BERR	RETRY	82
83	FC0	FC1	84
85	FC2	FC3	86
87	IRQ1	IRQ2	88
89	IRQ3	IRQ4	90
91	IRQ5	IRQ6	92
93	IRQ7	IACK	94
95	IACKIN	IACKOUT	96
97	BG0IN	BG0OUT	98
99	BG1IN	BG1OUT	100
101	BG2IN	BG2OUT	102
103	BG3IN	BG3OUT	104
105	BG4IN	BG4OUT	106
107	BR0	BR1	108
109	BR2	BR3	110
111	BR4	BBSY	112
113	BCLR	BREL	114
115	RESERVED	RESERVED	116
117	RESERVED	RESERVED	118
119	GND	GND	120
121	−12V	−12V	122
123	GND	GND	124
125	+12V	+12V	126
127	+12V	+12V	128
129	+5V	+5V	130
131	+5V	+5V	132
133	+5STDBY	+5STDBY	134
135	GND	GND	136
137	GND	GND	138
139	GND	GND	140

Fig. 20-2. Versabus P1 connector pinout designations.

Pin (Component Side)	Signal	Signal	Pin (Solder Side)
1	GND	GND	2
3	GND	GND	4
5	GND	GND	6
7	+5V	+5V	8
9	+5V	+5V	10
11	+12V	+12V	12
13	GND	GND	14
15	−12V	−12V	16
17	I/O NOT BUSED	I/O NOT BUSED	18
19	I/O NOT BUSED	I/O NOT BUSED	20
21	I/O NOT BUSED	I/O NOT BUSED	22
23	I/O NOT BUSED	I/O NOT BUSED	24
25	I/O NOT BUSED	I/O NOT BUSED	26
27	I/O NOT BUSED	I/O NOT BUSED	28
29	I/O NOT BUSED	I/O NOT BUSED	30
31	I/O NOT BUSED	I/O NOT BUSED	32
33	I/O NOT BUSED	I/O NOT BUSED	34
35	I/O NOT BUSED	I/O NOT BUSED	36
37	I/O NOT BUSED	I/O NOT BUSED	38
39	I/O NOT BUSED	I/O NOT BUSED	40
41	I/O NOT BUSED	I/O NOT BUSED	42
43	I/O NOT BUSED	I/O NOT BUSED	44
45	I/O NOT BUSED	I/O NOT BUSED	46
47	I/O NOT BUSED	I/O NOT BUSED	48
49	I/O NOT BUSED	I/O NOT BUSED	50
51	I/O NOT BUSED	I/O NOT BUSED	52
53	I/O NOT BUSED	I/O NOT BUSED	54
55	I/O NOT BUSED	I/O NOT BUSED	56
57	I/O NOT BUSED	I/O NOT BUSED	58
59	I/O NOT BUSED	I/O NOT BUSED	60
61	I/O NOT BUSED	I/O NOT BUSED	62
63	I/O NOT BUSED	I/O NOT BUSED	64
65	I/O NOT BUSED	I/O NOT BUSED	66
67	−15V	−15V	68
69	+15V	+15V	70
71	SERCOM 1	SERCOM 2	72
73	SERCOM 3	SERCOM 4	74
75	RESERVED	RESERVED	76
77	RESERVED	RESERVED	78
79	RESERVED	RESERVED	80
81	RESERVED	RESERVED	82
83	RESERVED	RESERVED	84
85	RESERVED	RESERVED	86
87	RESERVED	$\overline{\text{APARITY1}}$	88
89	$\overline{\text{A23}}$	$\overline{\text{A25}}$	90
91	$\overline{\text{A26}}$	$\overline{\text{A27}}$	92
93	$\overline{\text{A28}}$	$\overline{\text{A29}}$	94
95	$\overline{\text{A30}}$	$\overline{\text{A31}}$	96
97	GND	GND	98
99	$\overline{\text{DS2}}$	$\overline{\text{DS3}}$	100
101	GND	GND	102
103	$\overline{\text{DPARITY2}}$	$\overline{\text{DPARITY3}}$	104
105	$\overline{\text{D16}}$	$\overline{\text{D17}}$	106
107	$\overline{\text{D18}}$	$\overline{\text{D19}}$	108
109	$\overline{\text{D20}}$	$\overline{\text{D21}}$	110
111	$\overline{\text{D22}}$	$\overline{\text{D23}}$	112
113	$\overline{\text{D24}}$	$\overline{\text{D25}}$	114
115	$\overline{\text{D26}}$	$\overline{\text{D27}}$	116
117	$\overline{\text{D28}}$	$\overline{\text{D29}}$	118
119	$\overline{\text{D30}}$	$\overline{\text{D31}}$	120

Fig. 20-3. Versabus P2 connector pinout designations.

Connector and Pin No.	Signal	Definition
		master must receive the $\overline{\text{DTACK}}$ signal to complete the current operation.
P1 59	$\overline{\text{IOEN}}$	Input output enable—The $\overline{\text{IOEN}}$ is an active-low bus master output that is pulled low to indicate that address lines $\overline{\text{A1}}$–$\overline{\text{A15}}$ hold the address of an i/o device.
P1 66	$\overline{\text{RAMDIS}}$	RAM disable—The $\overline{\text{RAMDIS}}$ is an active-low line used to disable, or float, a section of the system's RAM. Use of the $\overline{\text{RAMDIS}}$ allows i/o overlays in RAM space. A block of i/o devices may, when $\overline{\text{RAMDIS}}$ is active, occupy the addresses normally used by RAM.
P1 64	$\overline{\text{ROMDIS}}$	ROM disable—The $\overline{\text{ROMDIS}}$ is an active-low line used to disable, or float, a section of the system's ROM. Use of the $\overline{\text{ROMDIS}}$ allows i/o overlays in ROM space. A block of i/o devices may, when $\overline{\text{ROMDIS}}$ is active, occupy the addresses normally used by ROM.
P1 74	$\overline{\text{SYSRESET}}$	System reset—The system reset is an active-low input signal that, when pulled low, initiates a system reset cycle.
P1 81	$\overline{\text{BERR}}$	Bus error—The $\overline{\text{BERR}}$ is an active-low bus master input pulled low by a peripheral when an unrecoverable error has occurred and the current bus cycle must be ended.
P1 82	$\overline{\text{RETRY}}$	Retry—The retry is an active-low bus master input pulled low by a peripheral when an error has occurred and the current operation must be attempted again.
P1 75	$\overline{\text{SYSPAUSE}}$	System pause—The system pause is an active-low signal that is used to cause all system processors to enter a wait state until the line is pulled high.
P1 76	SYSPACK	System pause acknowledge—The SYSPACK is an active-high open-collector line that indicates that all processors in the system have received the $\overline{\text{SYSPAUSE}}$ and have entered wait states.
P1 107–111	$\overline{\text{BR0}}$–$\overline{\text{BR4}}$	Bus request 0–4—The bus request lines are active-low, vectored bus request lines. BR0 has the highest priority. Each line may have several masters connected in a daisy chain configuration. When low, the bus request lines signal the current bus master that another master needs to use the bus. If the requesting master has a higher priority, the current master will be forced to relinquish control of the bus.
P1 112	$\overline{\text{BBSY}}$	Bus busy—The $\overline{\text{BBSY}}$ is an active-low bus master output that indicates the current master is using the bus and that lower priority masters must wait until the current master is finished before they can gain access to the bus. The current master pulls the BBSY line high as it begins its last cycle.
P1 97, 99, 101, 103, 105; 98, 100, 102, 104, 106	$\overline{\text{BG0IN}}$–$\overline{\text{BG4IN}}$ and $\overline{\text{BG0OUT}}$–$\overline{\text{BG4OUT}}$	Bus grant input 0–4 and bus grant output 0–4—The bus grant input 0–4 and output 0–4 lines form daisy chains that correspond to each of the bus request lines, $\overline{\text{BR0}}$–$\overline{\text{BR4}}$. Each board receives its signal on the bus grant input line from the next higher priority board's bus grant output line. Only one of the five bus grant daisy

Connector and Pin No.	Signal	Definition
		chains is activated by the processor at a time. If a board has requested use of the bus, by pulling its bus request line low, it also inhibits its bus grant output line, thereby disabling lower priority bus requests in that chain. The current master's bus grant output ripples down the chain selected on the basis of the priority of the bus request until it encounters the highest priority requesting device. That board may then gain control of the bus.
P1 113	$\overline{\text{BCLR}}$	Bus clear—the bus clear is an active-low bus master input used by the bus arbitrator to inform the current bus master that a master with a higher priority is requesting the bus and the current master must relinquish it at the end of the current operation.
P1 114	$\overline{\text{BREL}}$	Bus release—The bus release is an active-low bus master input used by a peripheral to inform the current bus master that it must release the bus within 16 bus cycles and it also causes the bus arbitrator to recognize only a request on $\overline{\text{BR0}}$.
P1 87–93	$\overline{\text{IRQ1}}$–$\overline{\text{IRQ7}}$	Interrupt request 1–7—The $\overline{\text{IRQ}}$ 1–7 lines are active-low bus master interrupt inputs. The $\overline{\text{IRQ}}$ lines are software maskable and are prioritized with $\overline{\text{IRQ7}}$ having the highest priority and $\overline{\text{IRQ1}}$ the lowest priority
P1 94	$\overline{\text{IACK}}$	Interrupt acknowledge—The interrupt acknowledge line is an active-low bus master output that is used by the master to acknowledge recognition of the interrupt request and ask that an interrupt vector be placed on data bus lines $\overline{\text{D0}}$–$\overline{\text{D7}}$. The bus master also outputs on $\overline{\text{A0}}$–$\overline{\text{A3}}$ the interrupt lines, $\overline{\text{IRQ1}}$–$\overline{\text{IRQ7}}$, to be serviced.
P1 95, 96	$\overline{\text{IACKIN}}$ $\overline{\text{IACKOUT}}$	Interrupt acknowledge in and interrupt acknowledge out—The $\overline{\text{IACKIN}}$ and $\overline{\text{IACKOUT}}$ lines are used to form a daisy chain interrupt priority scheme. The $\overline{\text{IACKIN}}$ line is fed by the next higher priority board's $\overline{\text{IACKOUT}}$. The chain is fed by the $\overline{\text{IACK}}$ line and the $\overline{\text{IACK}}$ signal is passed along the $\overline{\text{IACKIN}}$–$\overline{\text{IACKOUT}}$ chain until it reaches a board requesting an interrupt. That board compares the level to be serviced, which the bus master has placed on $\overline{\text{A0}}$–$\overline{\text{A3}}$, with its own level. If they match, it places the interrupt vector that points to its service routine on $\overline{\text{D0}}$–$\overline{\text{D7}}$ and inhibits its $\overline{\text{IACKOUT}}$, thereby preventing a lower priority board from responding. A level 7 board requests an interrupt by pulling $\overline{\text{IRQ7}}$ low. The bus master responds by pulling the $\overline{\text{IACK}}$ low. The $\overline{\text{IACK}}$ is fed to $\overline{\text{IACKIN}}$ on the highest priority board on each of the seven $\overline{\text{IRQ}}$ lines. The board requesting an interrupt can respond with a vector only if the bus master has output its $\overline{\text{IRQ}}$ line number on $\overline{\text{A0}}$–$\overline{\text{A3}}$ and if a higher priority board in its line has not requested service. The responding board places its vector on the bus and pulls the $\overline{\text{DTACK}}$ low.
P1 79	$\overline{\text{TEST}}$	Test—The $\overline{\text{TEST}}$ is an active-low line that forces all in-

Connector and Pin No.	Signal	Definition
		telligent boards on the bus to begin self-diagnosis routines.
P1 63	$\overline{\text{SECMAP}}$	Secondary map—The $\overline{\text{SECMAP}}$ is an active-low bus master output that is used to select devices in the secondary memory map. A system on the Versabus may have two memory maps with devices at the same locations. The bus master can enable devices in the second memory map and disable devices in the primary memory map by pulling $\overline{\text{SECMAP}}$ low.
P1 77	$\overline{\text{SECIEN}}$	Secondary interrupt enable—The $\overline{\text{SECIEN}}$ is an active-low bus master output used to enable interrupt and bus requests from devices in the secondary map.
P1 73	$\overline{\text{SECRESET}}$	Secondary reset—The $\overline{\text{SECRESET}}$ is an active-low bus master input that initiates reset of devices in the secondary map only.
P1 83–86	$\overline{\text{FC0}}$–$\overline{\text{FC3}}$	Function code 0–3—The function code lines are active-low bus master outputs that indicate the current state of the bus master. The code used on these lines is processor specific.
P1 70	SYSCLK	System clock—The system clock is a processor-independent 16-MHz signal that is generated by the bus master and may be used for timing operations.
P1 69	ACCLK	Ac clock—The ac clock line carries a signal that follows the power line frequency.
P2 17–66	I/O PIN	I/o pin—The i/o pins are lines that can be used for user-specific input and output functions.
P1 35, 115–118 P2 75–87	RESERVED	Reserved—The reserved lines are defined for use in 32-bit processor systems.
P1 78	$\overline{\text{ACFAIL}}$	Ac failure—The $\overline{\text{ACFAIL}}$ is an active-low bus master input that is pulled low when the primary ac input falls below the specified operating level.
P1 65	$\overline{\text{PDOWN}}$	Power down—The power down is an active-low bus master input that is pulled low when dc supply voltages fall below the specified operating levels.
P1 1, 2, 129–132 P2 7–10	+5V	+5 V—The +5-V line provides regulated +5 volts dc to system devices.
P1 133–134	+5STDBY	+5 V standby—The +5-V standby lines provides regulated +5 volts dc to devices that require battery backup when the +5-V supply is not operating.
P1 125–128 P2 11, 12	+12V	+12 V—The +12-V line provides regulated +12 volts dc to system devices.
P1 121–122 P2 15–16	−12V	−12 V—The −12-V line provides regulated −12 volts dc to system devices.
P2 69, 70	+15V	+15 V—The +15-V line provides regulated +15 volts dc to system devices.
P2 67, 68	−15V	−15 V—The −15-V line provides regulated −15 volts dc to system devices.
P1 3, 4, 23, 24, 27, 28, 31, 32, 61,	GND	Ground—The ground line is the common return for all system power supplies.

Connector and Pin No.	Signal	Definition
62, 67, 68, 71, 72, 119, 129, 123, 124, 135–140 P2 1–6, 13, 14, 97, 98, 101, 102	GND	Ground—The ground line is the common return for all system power supplies.

CHAPTER 21

The STD Bus

The STD bus was developed by Pro-Log and MOSTEK and is supported by more than 30 manufacturers. It differs from Intel's SBC and most other industrial buses in that it uses small, single-function boards that allow a system to contain only those functional blocks needed. A typical STD system, shown in Fig. 21-1, might consist of

Courtesy Pro-Log Corp.

Fig. 21-1. A typical STD system.

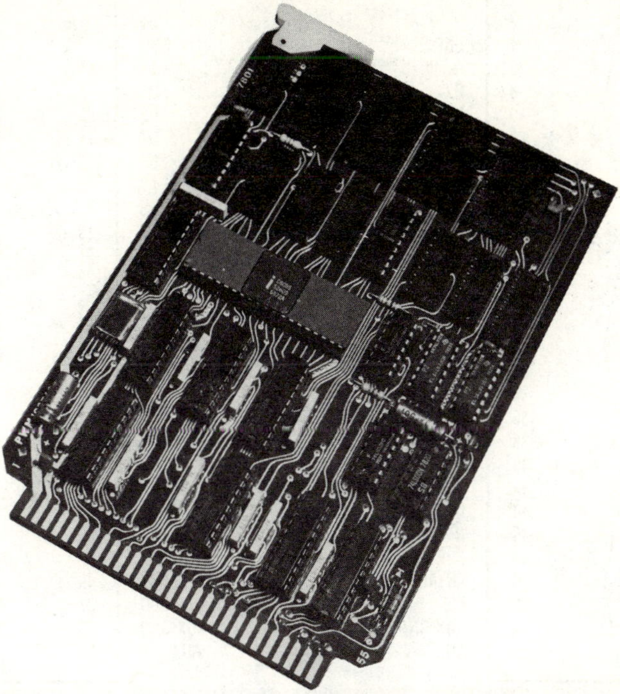

Courtesy Pro-Log Corp.

Fig. 21-2. An STD processor board.

a processor board, a ROM board, a RAM board, and as many parallel or serial i/o boards as needed to perform the required task. Dedicated systems are easily assembled using the small boards.

The STD bus accepts boards which are 6.5 inches (16.5 cm) long and 4.48 inches (11.3 cm) wide. An STD processor board is shown in Fig. 21-2. The STD uses a 56-pin, at 0.125 inch (0.317 cm), dual edge connector. The bus pinout designations are presented in Fig. 21-3. The STD is specified to be compatible with a variety of 8-bit processors and Z-80, 8085, 6800, and 6809 systems have been implemented on it. Because it can support several processors, STD lines are generally defined and allow processor specific definitions by processor board manufacturers.

STD SIGNAL DEFINITIONS

Pin No.	Signal	Definitions
7–14	D0–D7	Data bus—The bidirectional data bus lines carry data between the processor and peripherals.

2	+5V	+5V	1
4	GND	GND	3
6	−5V	−5V	5
8	D7	D3	7
10	D6	D2	9
12	D5	D1	11
14	D4	D0	13
16	A15	A7	15
18	A14	A6	17
20	A13	A5	19
22	A12	A4	21
24	A11	A3	23
26	A10	A2	25
28	A9	A1	27
30	A8	A0	29
32	$\overline{\text{RD}}$	$\overline{\text{WR}}$	31
34	$\overline{\text{MEMRQ}}$	$\overline{\text{IORQ}}$	33
36	$\overline{\text{MEMEX}}$	$\overline{\text{IOEXP}}$	35
38	$\overline{\text{MCSYNC}}$	$\overline{\text{REFRESH}}$	37
40	$\overline{\text{STATUS0}}$	$\overline{\text{STATUS1}}$	39
42	$\overline{\text{BUSRQ}}$	$\overline{\text{BUSAK}}$	41
44	$\overline{\text{INTRQ}}$	$\overline{\text{INTAK}}$	43
46	$\overline{\text{NMIRQ}}$	$\overline{\text{WAITRQ}}$	45
48	$\overline{\text{PBRESET}}$	$\overline{\text{SYSRESET}}$	47
50	$\overline{\text{CNTRL}}$	$\overline{\text{CLOCK}}$	49
52	PCI	PCO	51
54	AUX GND	AUX GND	53
56	AUX −V	AUX +V	55

Fig. 21-3. STD bus pinout designations.

Pin No.	Signal	Definitions
32	$\overline{\text{RD}}$	Read—The read line is an active-low processor output that signals the selected memory or i/o device to place data onto the data bus.
31	$\overline{\text{WR}}$	Write—The write line is an active-low processor output that signals the selected memory or i/o device that the data present on that data bus is valid and should be latched.
15–30	A0–A15	Address bus—The address bus is used by the processor to select the desired memory or i/o location.
34	$\overline{\text{MEMRQ}}$	Memory request—The memory request line is an active-low processor output that signals the presence of a valid memory address on the address bus. It is commonly used to enable memory devices and boards.
33	$\overline{\text{IORQ}}$	I/o request—The i/o request line is an active-low processor output that signals the presence of a valid i/o address on the address bus. It is commonly used to enable i/o devices and boards. The i/o request line may also be used with the $\overline{\text{M1}}$ line during the processor's response to an interrupt request to signal that an interrupt vector should be placed on the data bus by the device requesting service. Use of the $\overline{\text{IORQ}}$ for interrupt response occurs during the M1 period only.
45	$\overline{\text{WAITRQ}}$	Wait request—The wait request line is an active-low input to the processor used by slow memory or i/o devices to put the processor into a wait cycle.

Pin No.	Signal	Definitions
36	$\overline{\text{MEMEX}}$	Memory expansion—The memory expansion line is an active-low processor output that is used to expand memory addressing or enable memory devices. If unused in a system, this line is tied to ground.
35	$\overline{\text{IOEXP}}$	I/o expansion—The i/o expansion line is an active-low processor output that is used to expand i/o addressing or enable i/o devices. If unused in a system, this line is tied to ground.
38	$\overline{\text{MCSYNC}}$	Machine cycle sync—The machine cycle sync line is an active-low processor output that occurs once during each machine cycle. The specific function of this signal depends on the processor used. It is the SYNC on the 8080, the ALE on the 8085, the $\overline{\text{RD}}$, $\overline{\text{WR}}$, and $\overline{\text{INTAK}}$ ANDed on the Z-80, and the $\overline{\phi 2}$ on the 6800 and 6502.
39	$\overline{\text{STATUS 1}}$	Status 1—The status 1 line is an active-low processor output that indicates the beginning of an op-code fetch. The specific nature of this signal depends on the processor used. It is the M1 signal on the 8080, the S1 on the 8085, the M1 on the Z-80, the LIC on the 6809, and not generated by the 6800.
40	$\overline{\text{STATUS0}}$	Status 0—The status 0 line is an active-low processor output that is status 1 inverted on the 8085 and the valid memory address on the 6800. It is not available from the 8080, Z-80, or 6809.
37	$\overline{\text{REFRESH}}$	Refresh—The refresh line is an active-low line that signals the beginning of a refresh of dynamic memory cycle. The Z-80 provides the refresh signal. In systems not using the Z-80 the refresh must be supplied by another board.
47	$\overline{\text{SYSRESET}}$	System reset—The system reset line is an active-low output from the system reset circuitry to all system boards. It may be triggered by the power on process or by pushing the reset button.
48	$\overline{\text{PBRESET}}$	Push-button reset—The push-button reset line is an active-low input to the system reset circuitry from a manually activated reset button.
46	$\overline{\text{NMIRQ}}$	Nonmaskable interrupt—The nonmaskable interrupt request line is an active-low input to the processor's NMI or highest priority interrupt line.
44	$\overline{\text{INTRQ}}$	Interrupt request—The interrupt request line is an active-low input to processor's maskable interrupt input line.
43	$\overline{\text{INTAK}}$	Interrupt acknowledge—The interrupt acknowledge line is an active-low processor output that signals the device requesting an interrupt on the INTRQ line that the interrupt has been granted. In vectored interrupt systems it signals the requesting device to place an interrupt vector on the data bus.
51, 52	PCO, PCI	Priority chain out, priority chain in—These lines provide a position-dependent interrupt priority system when used in conjunction with the $\overline{\text{INTRQ}}$. (See the Apple II INT IN/OUT descriptions for an explanation of a position dependent interrupt priority system.)
42	$\overline{\text{BUSRQ}}$	Bus request—The bus request line is an active-low input that causes the processor to halt at the end of the current machine cycle and float the data, address, and control lines. In multiple processor systems this line may be an input, output, or bidirectional.
41	$\overline{\text{BUSAK}}$	Bus acknowledge—The bus acknowledge line is an active-low processor output used to indicate that the processor has floated the data, address, and control lines in response to a bus request input.
49	$\overline{\text{CLOCK}}$	Clock—The clock line is an active-low signal that may be used to synchronize system operations or other unspecified functions.

Pin No.	Signal	Definitions
50	$\overline{\text{CNTRL}}$	Control—The control line is an unspecified timing line that may be a processor input or output. It is typically used as an external clock input to the processor.
1, 2	+5V	+5 V—The +5-volt lines supply power to the logic circuits in the system.
5, 6	−5V	−5 V—The −5-volt lines supply power to the logic circuits in the system.
3, 4	GND	Ground—The ground line is the common power return for all logic circuits in the system.
55	AUX +V	Auxiliary positive voltage—The auxiliary positive voltage line supplies +12 volts to the system.
56	AUX −V	Auxiliary negative voltage—The auxiliary negative voltage line supplies −12 volts to the system.
53, 54	AUX GND	Auxiliary ground—The auxiliary ground line is the common return for the auxiliary supplies.

CHAPTER 22

The IEEE 488 Bus

The IEEE Standard 488-1978 defines a bus which is becoming very popular for electronic instrumentation. This standard has allowed instrument companies to build devices that can be easily used with different types of devices made by other companies. A system's hardware can be set up by simply plugging the devices together. Before the bus standard, an engineer designing an automatic test system would have to connect many kinds of cables with a wide variety of signals between the various instruments and controller(s).

There are three types of interface devices which can be connected to the bus: listeners, talkers, and controllers. Actually some devices are a combination of two or all of these, e.g., talker and listener. See Fig. 22-1 for examples. The bus standard allows a maximum of 15 devices and a maximum cable length of two times the number of devices (in meters) or 20 meters, whichever is less. A minimum configuration usually consists of one controller, one talker, and one listener (actually there could just be one talker and one listener). The bus allows multiple controllers which can transfer control back and forth.

Fig. 22-2 shows a bus controller with a keyboard for programming it. Once the controller is programmed, the keyboard can be removed. Fig. 22-3 shows two IEEE 488 instruments: a digital multimeter and a printer.

The bus is organized into three sets of signals: eight data lines, three data byte transfer control lines, and five general interface management lines. These are shown in Fig. 22-1 and the pinout des-

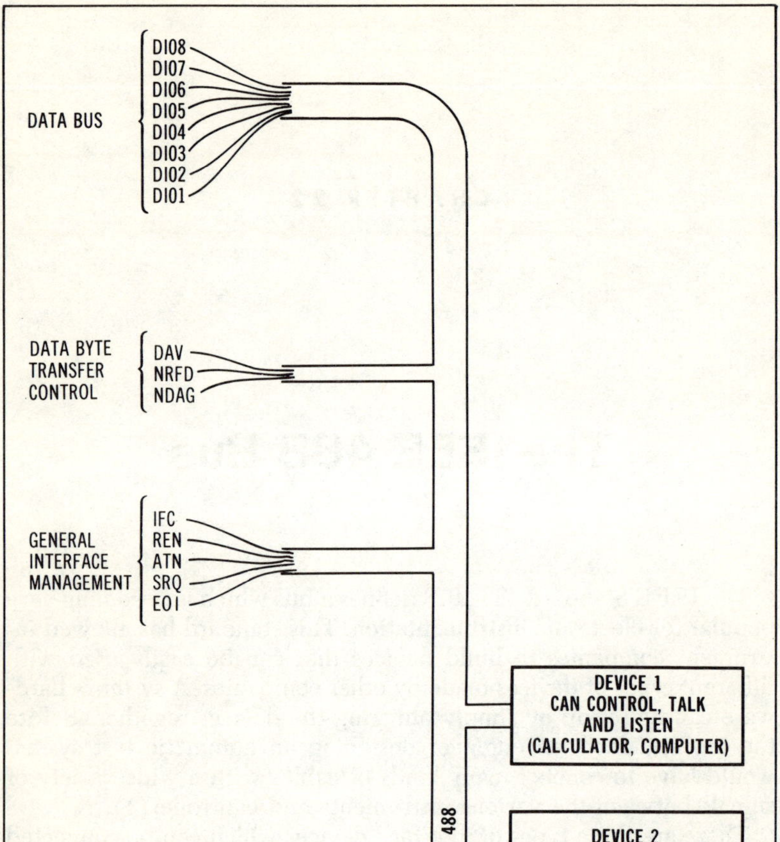

Fig. 22-1. An IEEE 488 system.

Courtesy Fluke, Inc.

Fig. 22-2. IEEE 488 bus controller.

ignations are given in Fig. 22-4. Note that there are several ground lines and a shield line, making a total of 24 lines. Data and interface messages are transmitted on the eight data lines in a bit-parallel, byte-serial, asynchronous, and bidirectional manner. The bus employs active-low (negative) logic where the more negative voltage is a true. The levels are typical TTL values.

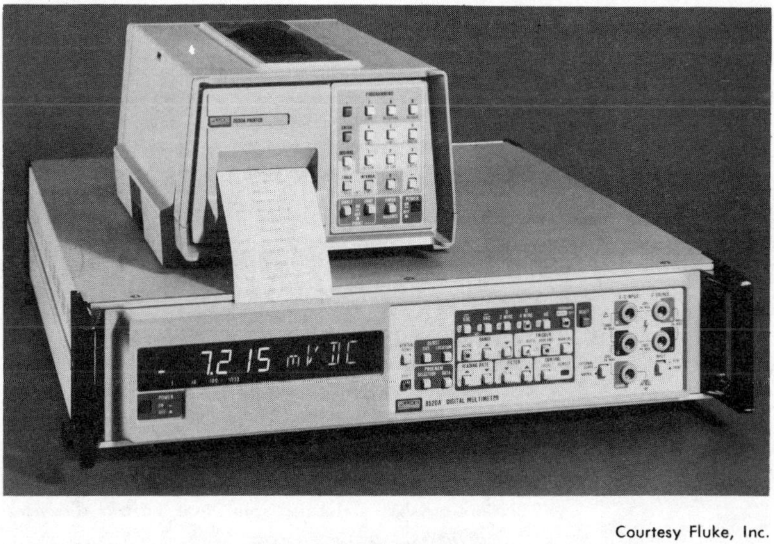

Courtesy Fluke, Inc.

Fig. 22-3. IEEE 488 instruments.

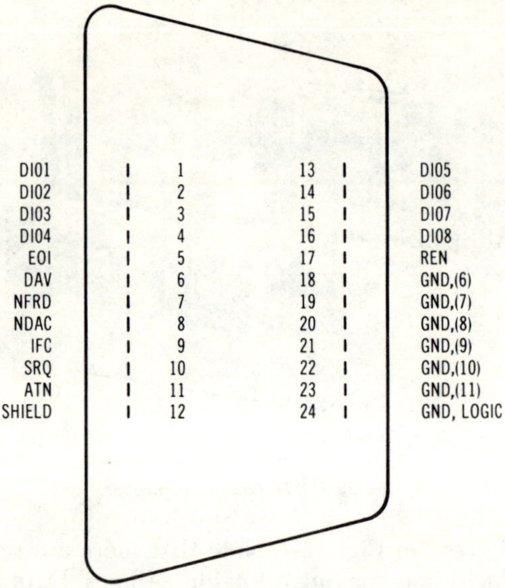

Fig. 22-4. IEEE 488 pinout designations.

IEEE 488 SIGNAL DEFINITIONS

Pin No.	Signal	Definition
1	DIO1	Data input output 1—DIO1–DIO8 are the eight bidirectional lines over which data is transmitted. Used to carry all interface and device dependent messages. The data on these lines may be in different forms at different times during usage e.g., 5-bit addresses, 8-bit binary numbers, two bcd values, ISO (ANSI) codes symbols and numbers (V for voltage), and one, two, or N bits of status information.
2	DIO2	Data input output 2—See pin 1.
3	DIO3	Data input output 3—See pin 1.
4	DIO4	Data input output 4—See pin 1.
5	EOI(24)	End or identify—Used by a talker to indicate the end of a multiple-byte transfer sequence, or, by a controller in conjunction with ATN pin 11, to execute a parallel polling sequence.
6	DAV	Data valid—Used to indicate the condition (availability and validity) of information on the DIO signal lines. Part of the handshake routine. See discussion for use of DAV, NRFD, and NDAC.
7	NRFD	Not ready for data—Used to indicate the condition of readiness of devices to accept data. Part of the handshake routine.
9	IFC	Interface clear—Used by a controller to place the interface system, portions of which are contained in all interconnected devices, in a known state.
10	SRQ	Service request—Used by a device to indicate the need for atten-

Pin No.	Signal	Definition
		tion and to request an interruption of the current sequence of events. When the line is pulled low by any system element to get the controller's attention, the controller responds at its discretion by conducting either a serial or parallel poll to identify the requesting device(s). See discussion on polls.
11	ATN	Attention—Used by a controller to specify how data on the DIO signal lines are to be interpreted and which devices must respond to the data. If this line is in the logic 1 (low) state, the information appearing on the DIO lines is interpreted as an interface control message; if it is in the logic 0 (high) state, the information is data. If both the ATN and EOI are true, a parallel polling sequence is initiated.
12	SHIELD	Shield—Ground line for cable shield. Connected to instrument's frames.
13	DIO5	Data input output 5—See pin 1.
14	DIO6	Data input output 6—See pin 1.
15	DIO7	Data input output 7—See pin 1.
16	DIO8	Data input output 8—See pin 1.
17	REN(24)	Remote enable—Used by a controller in conjunction with other messages to select between two alternate sources of programming data, either front-panel or interface. If REN is low (1) and the proper messages sent, the device will be remotely interface controlled.
18	GND, (6)	Ground—Ground wire paired with pin 6 (DAV). These may be a twisted pair.
19	GND, (7)	Ground—Ground wire paired with pin 7 (NFRD).
20	GND, (8)	Ground—Ground wire paired with pin 8 (NDAC).
21	GND, (9)	Ground—Ground wire paired with pin 9 (IFC).
22	GND, (10)	Ground—Ground wire paired with pin 10 (SRQ).
23	GND, (11)	Ground—Ground wire paired with pin 11 (ATN).
24	GND, LOGIC	Ground—Ground return for EOI (pin 5), REN (pin 17), and DIO lines.

The IEEE 488 connector is shown in Fig. 22-5. This and associated cables can be purchased from several manufacturers or suppliers. Cables typically have both a plug and receptacle mounted together (Fig. 22-6) so that several cables may be connected in any desired configuration.

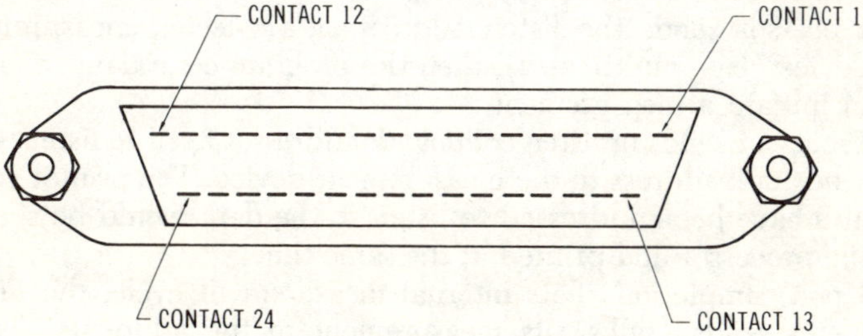

Fig. 22-5. IEEE 488 connector, showing contacts.

Fig. 22-7 shows a typical system using the IEEE 488 bus. The sequence given is to demonstrate a possible utilization of the bus.

1. Processor sends IFC true, thereby initializing the interface.
2. Processor sends DCL measure true, causing all devices to be set to a predefined state.
3. Processor sends the listen address of the power supply followed by data to program its output(s).
4. Processor sends the unlisten command, followed by the listen address for the function generator, followed by data to program it.

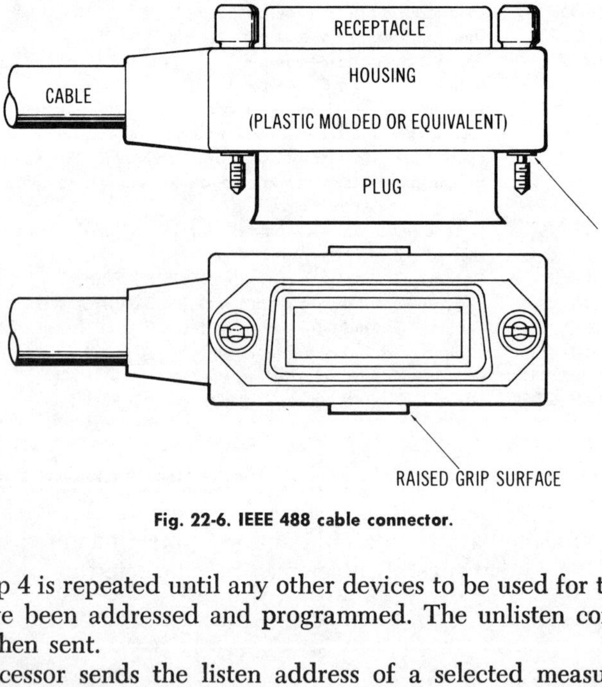

Fig. 22-6. IEEE 488 cable connector.

5. Step 4 is repeated until any other devices to be used for this test have been addressed and programmed. The unlisten command is then sent.
6. Processor sends the listen address of a selected measurement device (e.g., multimeter), then the program code data required to initiate a measurement.
7. Processor sends unlisten command, addresses itself to listen, then sends talk address to the measurement device. The printer could also have been addressed to listen so the data would be sent to the processor and printed at the same time.
8. Upon completion of its internal measurement cycle, the multimeter sends (talks) its measurement to the addressed listener(s).

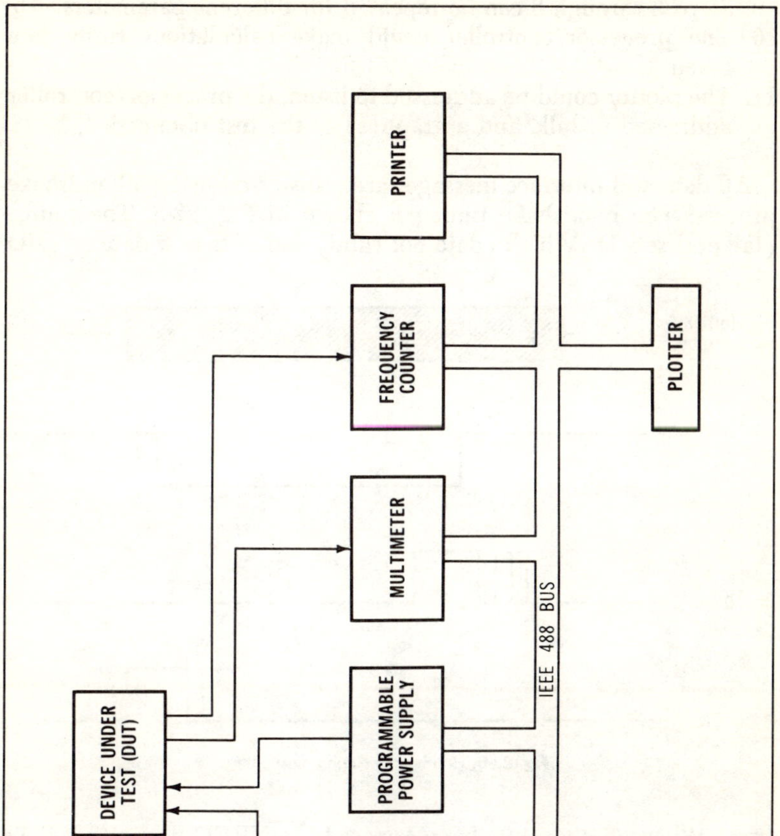

Fig. 22-7. Typical IEEE 488 system.

9. Steps 3 through 8 can be repeated for different parameters.
10. The processor/controller could make calculations from data taken.
11. The plotter could be addressed to listen, the processor/controller addressed to talk, and a graph(s) of the test obtained.

All data and interface messages are transferred using a handshake process. The handshake timing is shown in Fig. 22-8. The source (talkers) sets DAV high (data not valid) and puts new data on DIO

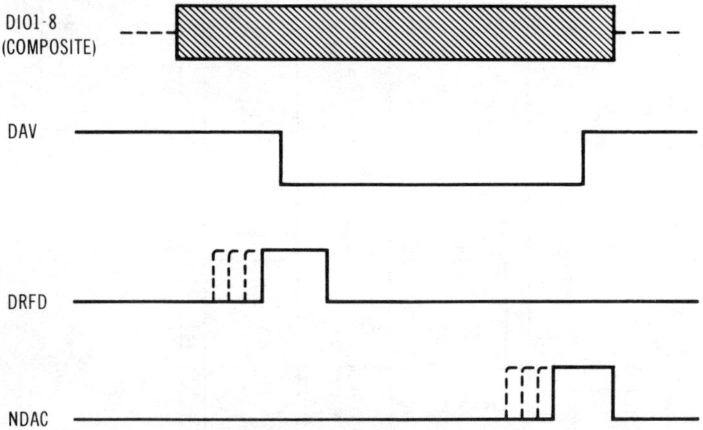

Fig. 22-8. Handshake timing sequence.

lines. When all acceptors have released the NRFD, indicating their readiness for new data, the source sets DAV low (data valid) and holds it and the data stable until all acceptors release NDAC. This means all devices listening have accepted the data byte. The source now sets DAV high and the acceptors pull NDAC low. The cycle repeats for each data (or interface message) byte transferred.

Fig. 22-9 shows the interface messages in terms of the 7-bit code. The figure has two columns for each number. The left is the ISO code. It is selected by the ATN line. If ATN = 1, the 7 bits are interface messages, and if ATN = 0, the data lines contain data which can be in ISO form.

Some examples may clarify the messages. Suppose the controller output ATN = 1 and DIO = X0101011. This is recognized as an interface message (ATN = 1) and is in the 010 column, which means that it is telling a device (01011) to listen. Suppose the controller output ATN=1 and DIO = X1011111. From Fig. 22-8 this gives the UNT (untalk) message, which removes any device from the talker

(SENT AND RECEIVED WITH ATN=1)

b_7 b_6 b_5 BITS b_4	b_3	b_2	b_1	COLUMN → ROW ↓	0_{0_0}	MSG ①	0_{0_1}	MSG	0_{1_0}	MSG	0_{1_1}	MSG	1_{0_0}	MSG	1_{0_1}	MSG	1_{1_0}	MSG	1_{1_1}	MSG
					0		1		2		3		4		5		6		7	
0	0	0	0	0	NUL		DLE		SP	MLA ASSIGNED TO DEVICE	0	MLA ASSIGNED TO DEVICE	@	MTA ASSIGNED TO DEVICE	P	MTA ASSIGNED TO DEVICE		MEANING DEFINED BY PCG CODE	p	MEANING DEFINED BY PCG CODE
0	0	0	1	1	SOH	GTL	DC1	LLO	!		1		A		Q		a		q	
0	0	1	0	2	STX		DC2		'		2		B		R		b		r	
0	0	1	1	3	ETX		DC3		#		3		C		S		c		s	
0	1	0	0	4	EOT	SDC	DC4	DCL	$		4		D		T		d		t	
0	1	0	1	5	ENQ	PPC ③	NAK	PPU	%		5		E		U		e		u	
0	1	1	0	6	ACK		SYN		&		6		F		V		f		v	
0	1	1	1	7	BEL		ETB				7		G		W		g		w	
1	0	0	0	8	BS	GET	CAN	SPE	(		8		H		X		h		x	
1	0	0	1	9	HT	TCT	EM	SPD	)		9		I		Y		i		y	
1	0	1	0	10	VT		SUB		*		:		J		Z		j		z	
1	0	1	1	11	FF		ESC		+		;		K		[		k		{	
1	1	0	0	12	CR		FS		.		<		L		\		l		¦	
1	1	0	1	13	SO		GS		—		=		M		]		m		}	
1	1	1	0	14	S1		RS		.		>		N		^		n		~	
1	1	1	1	15	GTL		US		/		?	UNL	O		___	UNIT	o		DEL	

ADDRESSED COMMAND GROUP (ACG)

UNIVERSAL COMMAND GROUP (UCG)

LISTEN ADDRESS GROUP (LAG)

TALK ADDRESS GROUP (TAG)

PRIMARY COMMAND GROUP (PCG)

SECONDARY COMMAND GROUP (SCG)

NOTE:

① MSG = INTERFACE MESSAGE

② b_1 = DI01...by = DI07

③ REQUIRES SECONDARY COMMAND

④ DENSE SUBSET (COLUMNS 2 THROUGH 5)

Fig. 22-9. Multiline interface messages: ISO–7 bit code representation.

mode. One more example: The controller outputs ATN = 1 and DIO = X0010100, which is the DCL (device clear) message. This then clears (reset) any devices which are in the listener mode.

An IEEE 488 device has a set of functions selected from Table 22-1. Many devices only need a few of these functions and the designer is free to select these. The symbols are often printed next to the plug on the device, e.g. SH1, AH1, T2, L1, SR1, RL2, PP2, DC1, DT1, E0, E1.

Table 22-1. Interface Function Repertoire

Function	Symbol	Capabilities
Source handshake	SH	0, 1
Acceptor handshake	AH	0, 1
Talker or extended talker	T or TE	0–8
Listener or extended listener	L or LE	0–4
Service request	SR	0, 1
Remote local	RL	0, 1, 2
Parallel poll	PP	0, 1, 2
Device clear	DC	0, 1, 2
Device trigger	DT	0, 1
Controller	C	0–28

The symbol with zero means that the device has no capabilities in that interface function. Some functions have only two levels (0, 1) whereas others have several levels of capabilities. The IEEE 488 specifications can be consulted if the complete capabilities of each function are needed. The E1 at the end of the example designates open-collector devices whereas an E2 would designate three-state devices, which are used for faster data transfer rates.

The IEEE 488 allows both serial and parallel polling to determine devices requesting service.

A serial poll is initiated by one or more devices with the talker function asserting the SRQ bus line. The controller then sequentially addresses the device on the bus in whatever order of priority desired. The currently addressed device would then put a status byte on the DIO lines. If that particular device has asserted the SRQ line, it would also put a true (1) on DIO7 and other status information on DIO0–DIO6 and DIO8. If DIO7 were false (0), then the controller would proceed with the next device and so on. Only devices which have received the serial poll enable (SPE) message would respond since they should power-up disabled. Devices can be made to not respond by sending the serial poll disable (SPD) message. In serial polling, the handshake is used.

Parallel polling is done by the bus controller. In order for a device to respond, it must be sent the parallel poll configure (PPC) message and the parallel poll enable (PPE) message. The PPE message is:

DIO							
8	7	6	5	4	3	2	1
Y	1	1	0	S	P3	P2	P1

where Y is a don't care, S determines the sense of the response (S=1 means response is 1), and P3, P2, and P1 determine which DIO line that device responds on (000 = DIO1, 111 = DIO8). The controller asks for a parallel poll by asserting the ATN and EOI lines. No handshake is used. The devices' enable would then assert their configured data lines if requesting service and the controller would input the data byte. Up to eight devices can be polled at once. If more than one device is configured for the same data line, more than eight devices can be checked. Some additional polling scheme would be required though. Devices can be made to not respond by either the parallel poll unconfigure (PPU) message or the parallel poll disable (PPD) message. These are slightly different and the IEEE standard should be consulted if necessary.

CHAPTER 23

Benton Harbor Bus to S-100 Bus

A Heath Company H8 system would become more flexible if the wide variety of available S-100-compatible boards could be connected to it. An H8 system owner might also save money by using the competitively priced S-100 boards. This chapter will discuss the conversion of Benton Harbor bus (BHB) signals to S-100 bus signals.

To accomplish a BHB to S-100 bus conversion, a conversion board is needed. It would fit between the BHB and S-100 buses as shown in Fig. 23-1. The easiest method would be to make the conversion board plug into an S-100 motherboard and to have jumpers run to a connector for the BHB motherboard. The conversion board could be either a printed-circuit or wire-wrapped board.

The address bus A0–A15 lines can be converted fairly easily as shown in Fig. 23-2. They only need to be connected to the S-100 lines through 16 inverting three-state buffers. They are also connected to the address disable $\overline{\text{ADD DSB}}$ line through an inverter. If DMA or other bus control devices are not used, the lines could simply be connected through 16 inverters and the $\overline{\text{ADD DSB}}$ line could be omitted.

The BHB data lines $\overline{\text{D0}}$–$\overline{\text{D7}}$ must be inverted and separated into data-in (DI0–DI7) and data-out (DO0–DO7) lines on the S-100 bus. Fig. 23-3 shows a possible conversion of the data bus that is accomplished using 16 inverting three-state buffers. The data output lines can be put in a high-impedance state by $\overline{\text{DO DSB}}$. However, if DMA is not used, $\overline{\text{DO DSB}}$ can be omitted and eight inverters used instead. Either type of read signal (IOR or MEMR) will allow the data input signals to pass inverted to the BHB data lines.

The S-100 status lines are obtained as shown in Fig. 23-4. These lines can be disabled by connecting $\overline{\text{STAT DSB}}$ to the three-state buffers. Again, without DMA, the $\overline{\text{STAT DSB}}$ and the buffers can

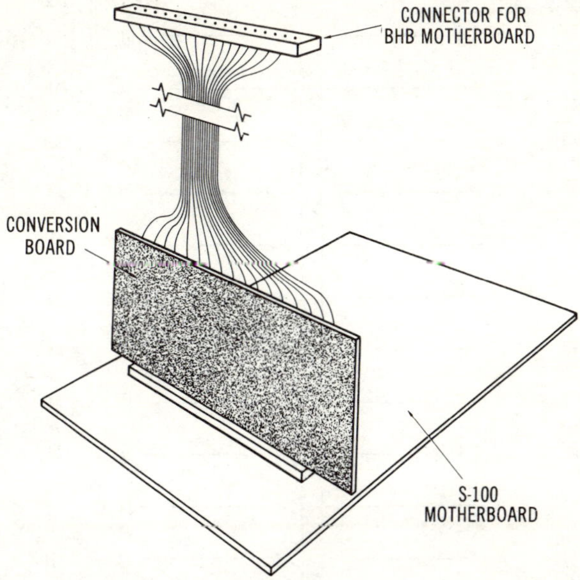

Fig. 23-1. Connection of an S-100 motherboard to a Bentor Harbor motherboard via a conversion board.

be omitted. On the S-100 bus, status signals are latched onto the bus and remain there until the next machine cycle. In the conversion system shown, the timing will be quite different since these signals are not available on the BHB. The derived lines should have the necessary active level at the proper time for most boards used. The two signals SHLTA and SSTACK are not obtainable from the BHB but should not restrict the use of memory boards. Also shown in Fig. 23-4 is the MWRITE signal. The interrupt acknowledge signal, SINTA, is not obtainable from the BHB but could be jumpered from the H8 CPU board if required.

Fig. 23-5 shows a possible conversion method for obtaining the command/control signals. The lines are shown with three-state buffers which can be omitted if DMA does not require the command/control disable $\overline{\text{C/C DSB}}$. Signals PSYNC, PWAIT, and PINTE are not obtainable from the BHB; but, if required, they exist on the H8 CPU board and could be jumpered.

A conversion method for the ready and interrupt lines is shown in Fig. 23-6. If either of the S-100 ready lines $\overline{XRDY}$ or $\overline{PRDY}$ goes high, the processor will enter a wait state. The BHB has priority interrupts, with INT7 being the highest. Interrupts INT1 and INT2

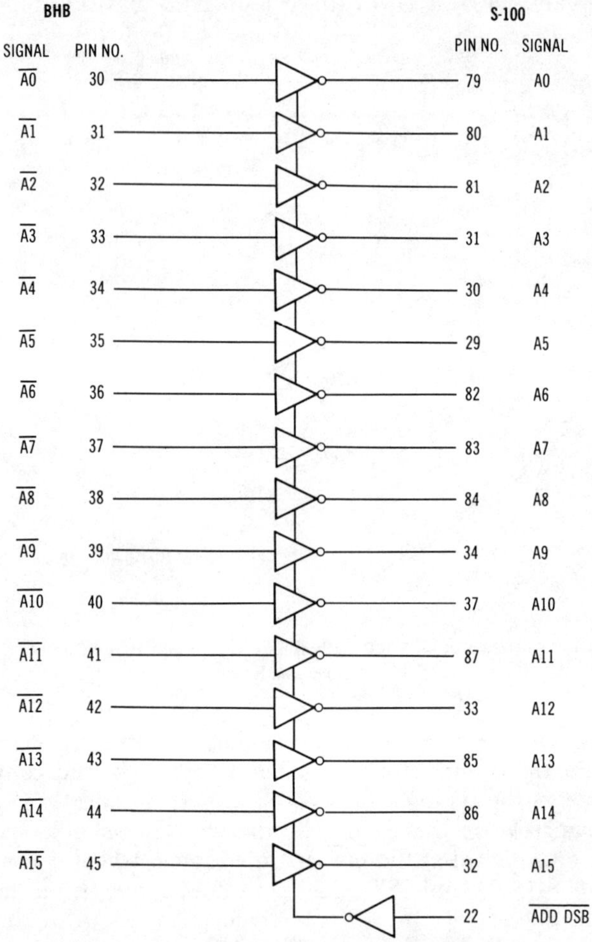

Fig. 23-2. Simple method used to convert the address bus lines.

on the BHB are dedicated to the H8 front panel. If the front panel is not being used, however, jumpers can be placed on the H8 CPU board to allow other devices to use INT1 and INT2. One method of handling interrupts is with the H8 priority system shown in Fig. 23-6. This allows five (seven without front panel) priority interrupts. However, if interrupts from peripherals occur on the BHB, the number of lines available will be reduced accordingly. If more interrupts are required, an interrupt-controller board on the S-100 bus could accept VI0–VI7 signals and generate only a $\overline{\text{PINT}}$ interrupt signal which would be connected through an inverter to an unused interrupt on the BHB. If an S-100 board has VI0–VI7 active high, then inverters would have to be added to the lines shown in Fig. 23-6.

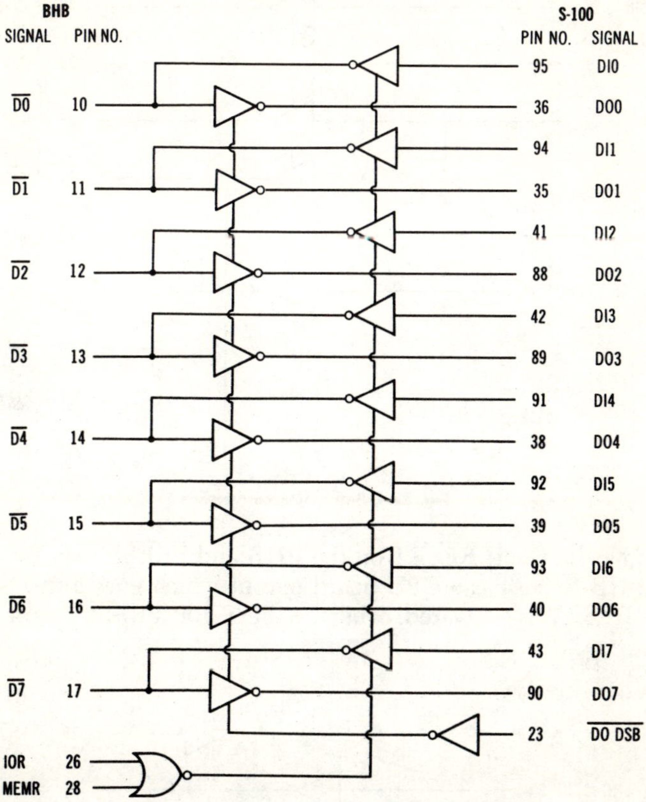

Fig. 23-3. A possible data bus conversion method.

The remaining bus conversions are shown in Fig. 23-7. The power supplies and ground are straightforward. Note the higher voltage levels, however. This is not a problem since the on-board regulators derive the proper voltages from these voltages. The BHB $\phi 2$ clock signal provides both $\overline{\text{CLOCK}}$ and $\phi 2$ through an inverter. The clock can be used as $\phi 1$ on the S-100 bus in most cases but, if true $\phi 1$ is required, an additional nonoverlapping clock circuit will be necessary. The BHB $\overline{\text{HOLD}}$ connects directly to the S-100 $\overline{\text{PHOLD}}$. The BHB $\overline{\text{ROM DISABLE}}$ does not convert to an S-100 signal. Both

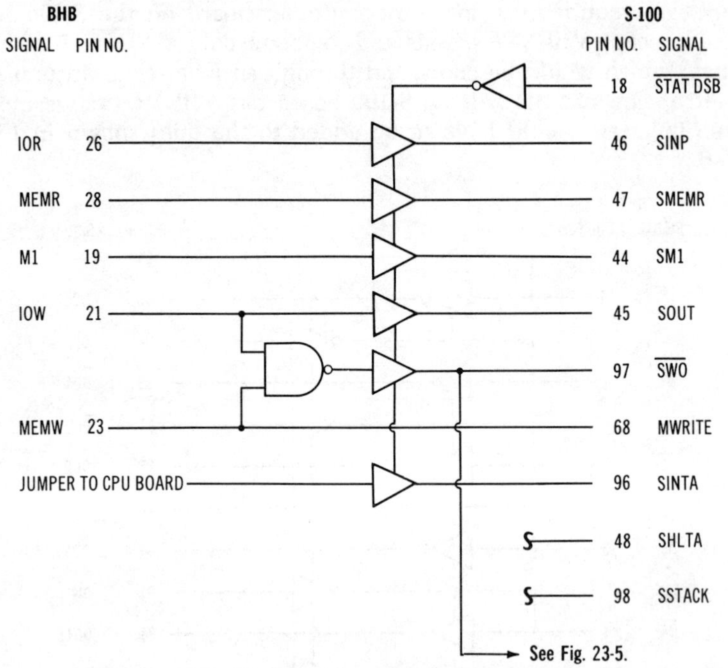

Fig. 23-4. Status line conversion.

buses have a reset: $\overline{\text{RESET}}$ on the BHB and $\overline{\text{PRESET}}$ on the S-100. The BHB $\overline{\text{RESET}}$ is a CPU board output which goes low when the CPU RESET is activated, whereas the S-100 $\overline{\text{PRESET}}$ is a signal

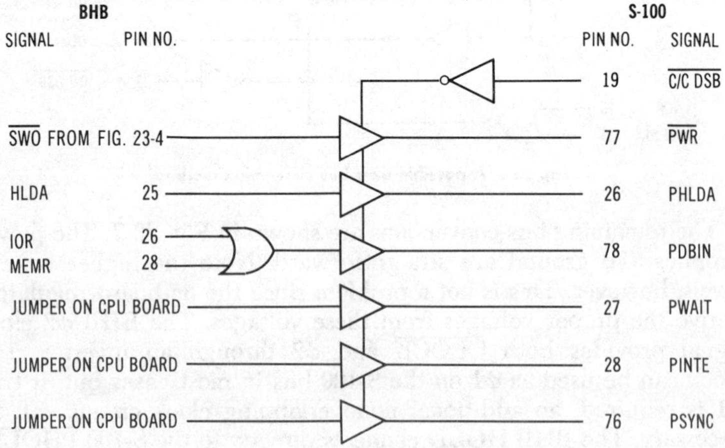

Fig. 23-5. Command/control line conversion.

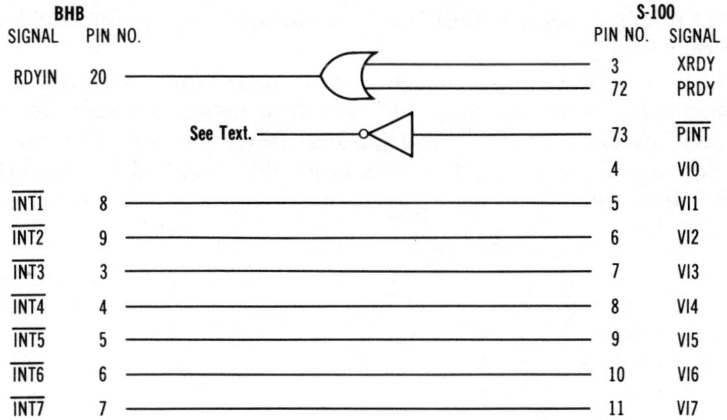

Fig. 23-6. A method that can be used to convert ready and interrupt lines.

to reset the CPU. Connecting these lines together would allow a BHB CPU RESET to also reset any other boards connected to the S-100 bus. Jumpering the S-100 $\overline{\text{PRESET}}$ to $\overline{\text{RESIN}}$ on either the H8 CPU board or the front panel would allow boards on the S-100 bus to reset the CPU.

S-100 signals not accounted for are PROT (pin 70), RUN (pin 71), UNPORT (pin 20), SSWI (pin 53), PS (pin 69), SS (pin 21), $\overline{\text{EXT CLR}}$ (pin 54), $\overline{\text{PHANTOM}}$ (pin 67), and POC (pin 99). Some of these lines may be implemented with front-panel switches and indicators. Obtaining others may require sophisticated circuitry.

Any effort to use S-100 boards from the BHB, by using a conversion board as discussed, should be preceded by the gathering of as much information as possible about the boards. The required S-100 signals should be determined and compared to either what is available on the conversion board or what can be created elsewhere. Any

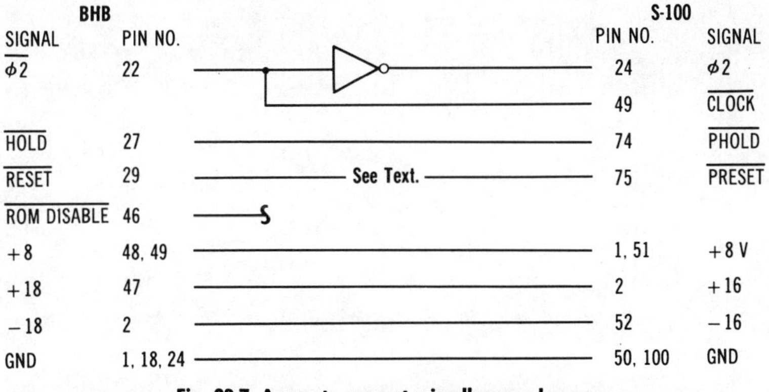

Fig. 23-7. A way to convert miscellaneous buses.

strict timing of signals must be known to determine if any problems lie therein.

If it is desired to use a conversion board which would permit using BHB boards on an S-100 system, it could be easily accomplished by deriving the lines as on the H8 CPU board. This can be done since the S-100 system was originally designed for the 8080 CPU as is the H8.

CHAPTER 24

TRS-80 Model I Bus to S-100 Bus

The owner of a TRS-80 Model I system might like to use S-100 bus boards because of their lower cost and wider variety. Therefore, this chapter discusses the conversion of signals from the TRS-80 bus to the S-100 bus. A conversion board, either printed-circuit or wire-wrapped, is needed. It should be designed to plug into an S-100 motherboard and should be cabled to a connector which fits the TRS-80 expansion edge card.

The address bus A0–A15 can be connected directly from the TRS-80 Model I bus to the S-100 bus. If several S-100 boards are used, it might be preferable to buffer the address lines on the conversion board. However, if DMA is used, the buffers should be three-state and be enabled by the $\overline{\text{TEST}}$ (pin 23) line on the TRS-80 Model I bus (connected through an inverter).

A data bus D0–D7 conversion is shown in Fig. 24-1. The data lines are divided into data in (DI0–DI7) and data out (DO0–DO7) for the S-100 bus. Both the data-in and data-out lines are buffered. The $\overline{\text{TEST}}$ line on the TRS-80 Model I bus three-state buffers the data out for DMA operations. The $\overline{\text{RD}}$ and $\overline{\text{IN}}$ lines enable the data in for read or input operations. IF DMA is not needed, the data-out buffers can be omitted *provided* there is not an excessive number of boards on the bus.

The ground lines on the TRS-80 Model I bus (pins 8, 29, and 37) are connected to the ground lines on the S-100 bus (pins 50 and 100). However, the TRS-80 Model I bus has only a +5-V power supply, whereas the S-100 bus has a +8-V and a ±16-V power supply. There-

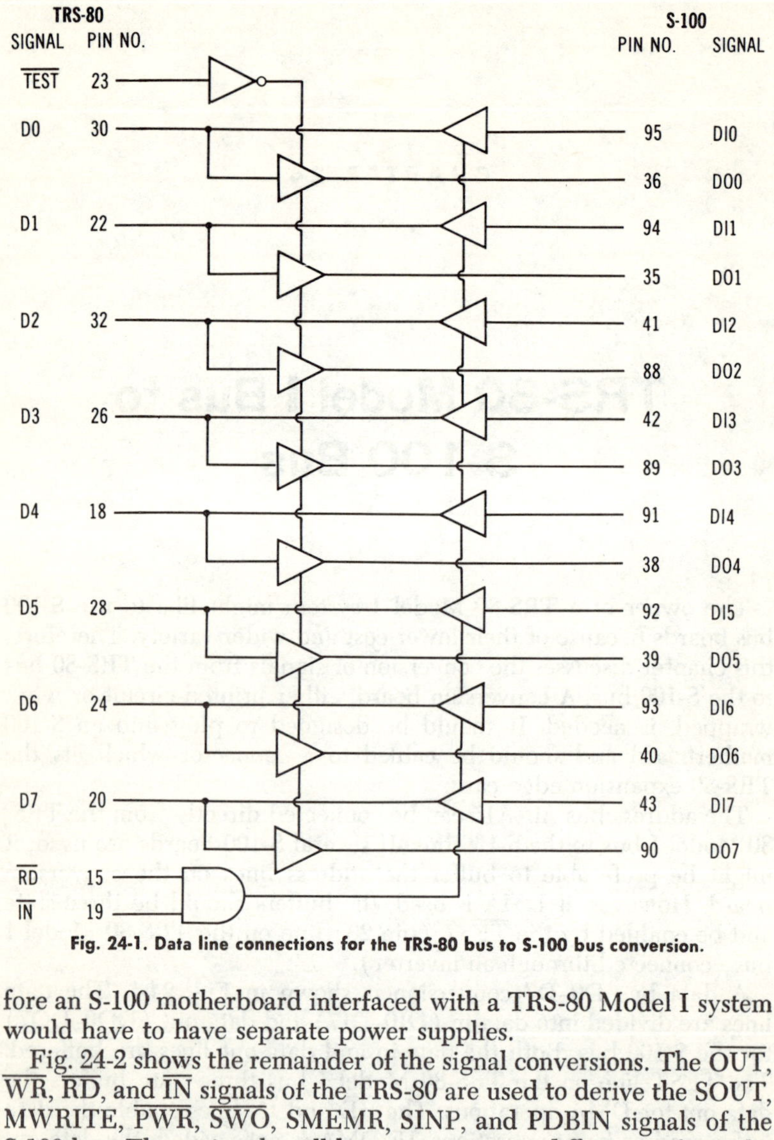

Fig. 24-1. Data line connections for the TRS-80 bus to S-100 bus conversion.

fore an S-100 motherboard interfaced with a TRS-80 Model I system would have to have separate power supplies.

Fig. 24-2 shows the remainder of the signal conversions. The OUT, WR, RD, and IN signals of the TRS-80 are used to derive the SOUT, MWRITE, PWR, SWO, SMEMR, SINP, and PDBIN signals of the S-100 bus. These signals will have some timing differences from the true S-100 signals but will work on most boards. The TRS-80 Model I bus has only one interrupt input—INT. If vectored interrupts are desired, an interrupt controller board can be used on the S-100 bus.

Any device desiring DMA on the S-100 bus activates PHOLD and the various disable lines. In the conversion, these lines are ORed to provide the TEST signal on the TRS-80 Model I bus. This puts all

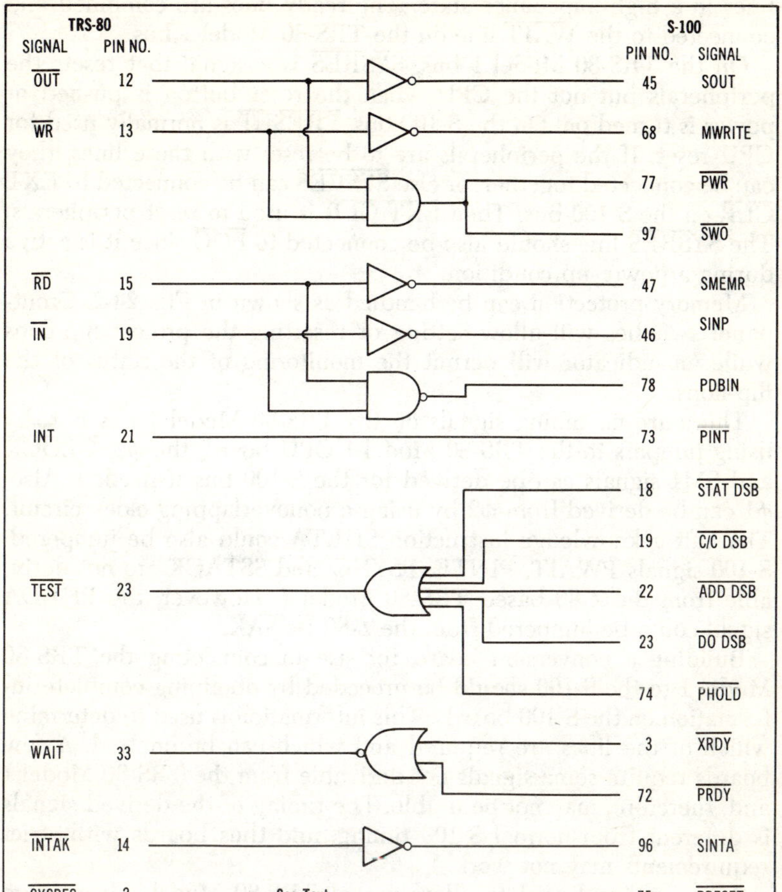

Fig. 24-2. Line connections for the signal conversions not shown in Fig. 24-1.

lines in a high-impedance state. The ready lines are combined and connected to the $\overline{\text{WAIT}}$ line on the TRS-80 Model I bus.

On the TRS-80 Model I bus, $\overline{\text{SYSRES}}$ is a signal that resets the peripherals but not the CPU when the reset button is pushed or power is turned on. On the S-100 bus, $\overline{\text{PRESET}}$ is normally used for CPU reset. If the peripherals are to be reset with these lines, they can be connected together, or else $\overline{\text{SYSRES}}$ can be connected to EXT $\overline{\text{CLR}}$ on the S-100 bus. Then $\overline{\text{EXT CLR}}$ is used to reset peripherals. The $\overline{\text{SYSRES}}$ line should also be connected to $\overline{\text{POC}}$ since it is active during a power-up condition.

Memory protection can be handled as shown in Fig. 24-2. Front-panel switches will allow setting or resetting the protect flip-flops while an indicator will permit the monitoring of the status of the flip-flops.

There are no timing signals on the TRS-80 Model I bus but, by using jumpers to the TRS-80 Model I CPU board, the $\phi 2$, $\overline{\text{CLOCK}}$ and SM1 signals can be derived for the S-100 bus if needed. Also, $\phi 1$ can be derived from $\phi 2$ by using a nonoverlapping clock circuit. The halt acknowledge instruction SHLTA could also be jumpered. S-100 signals PWAIT, PINTE, PSYNC, and SSTACK are not derivable from the Z-80 based TRS-80 Model I. However, the PHLDA signal could be jumpered from the Z-80 $\overline{\text{BUSAK}}$.

Building a conversion board for use in connecting the TRS-80 Model I to the S-100 should be preceded by obtaining complete information on the S-100 boards. This information is used to determine which of the lines are required and which can be omitted. A few boards require some signals not derivable from the TRS-80 Model I and, therefore, may not be usable. The timing of the derived signals is different from normal S-100 timing, and thus boards with strict requirements may not work.

A conversion board to allow use of TRS-80 Model I boards on S-100 systems is possible as most of the signals on the TRS-80 bus are readily derived. However, the $\overline{\text{CAS}}$ and MUX signals which are used for 16-pin dynamic memory are not available and, therefore, an additional timing circuit to generate these signals must be included. See Chapter 5 and Fig. 5-6 for a description of a TRS-80 Model I circuit that will generate these signals.

CHAPTER 25

6502/6800 to S-100 Conversion

Although 6502/6800-based systems have been quite popular, relatively few expansion boards have been produced for them. Only the SS-50 bus offers users a variety of boards. However, use of the plentiful S-100 boards on 6502/6800 systems is possible via a bus interface board.

The KIMSI, shown in Figs. 25-1 and 25-2, is a good, commercially produced KIM to S-100 interface board that will be used to illustrate the needed signal conversions. The KIMSI interface board plugs onto the KIM expansion connector and provides an eight-slot S-100 motherboard.

Three types of signals (address, data, and control) must be transferred between the buses. Since the KIMSI is not provided on-board, all KIM signals must be buffered. Both systems use 16-line address buses, so the A0–A15 signals of the 6502 translate directly into the S-100 address lines as shown in Fig. 25-3. The 74367 buffers provide the needed drive.

The bidirectional data bus of the 6502 must be divided into data in and data out lines as shown in Fig. 25-4. Data out and data in are both connected to the D0–D7 lines of the 6502 through three-state buffers. The data out buffers are controlled by the $\overline{R/W}$ line of the KIM and the data in buffers are controlled by the R/W line of the KIM. The possibility of input and output data appearing on the D0–D7 lines simultaneously is, therefore, prevented.

The $\phi 2$ line of the 6502 is inverted to become the S-100 $\phi 2$ line and can be jumpered to K$\phi 2$. The SYNC signal of the 6502 which

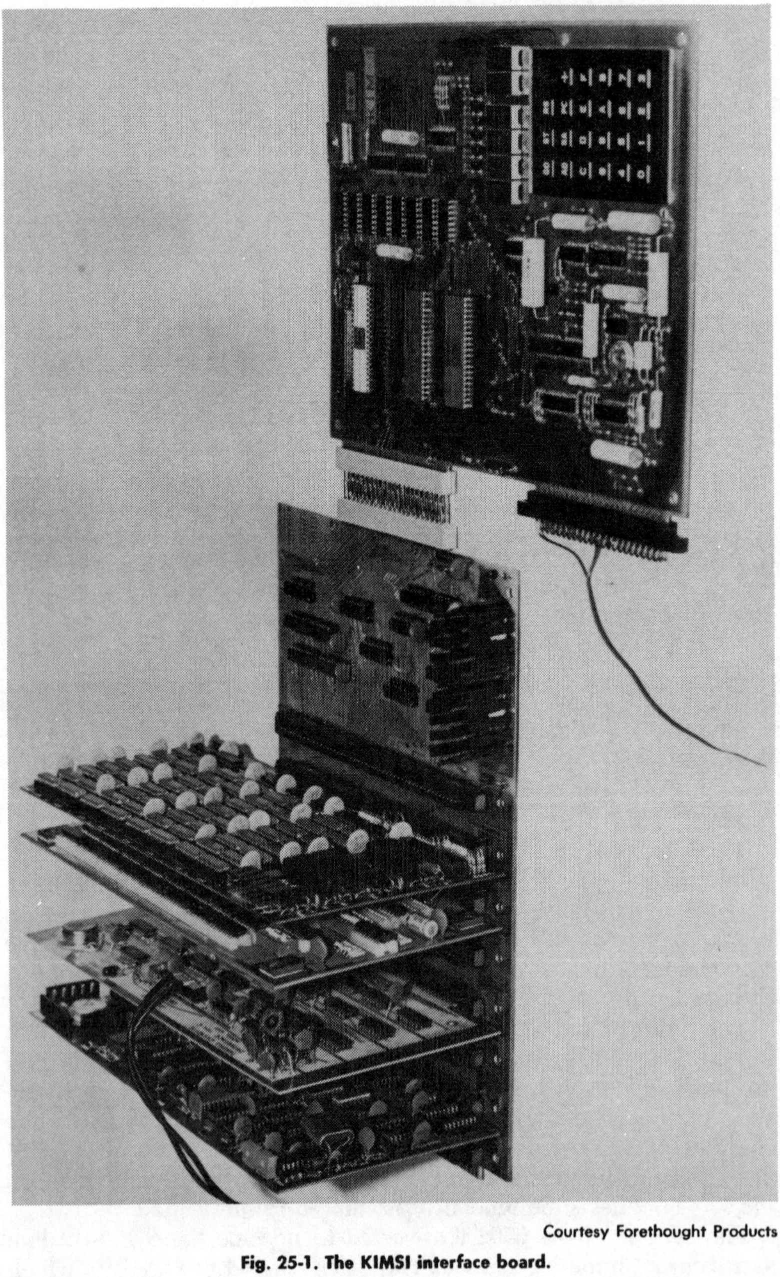

Courtesy Forethought Products

Fig. 25-1. The KIMSI interface board.

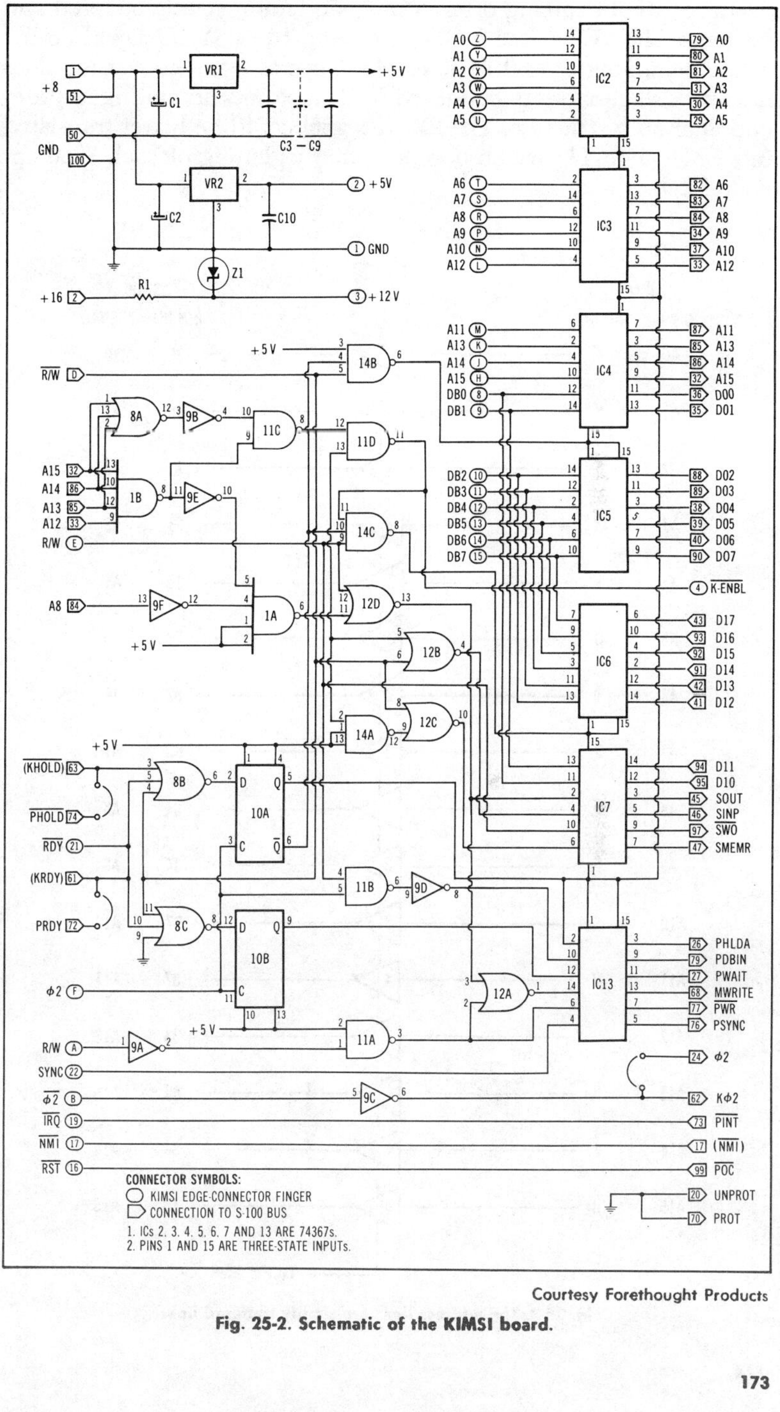

Courtesy Forethought Products

Fig. 25-2. Schematic of the KIMSI board.

occurs at the beginning of each op-code fetch cycle is buffered and becomes PSYNC for the S-100. However, the PSYNC signal occurs at the beginning of each 8080 machine cycle so the conversion does not work since more than one PSYNC signal occurs during the execution of an instruction. The SYNC signal could be better translated into SM1 (line 44) which occurs at the beginning of each 8080 op-code fetch.

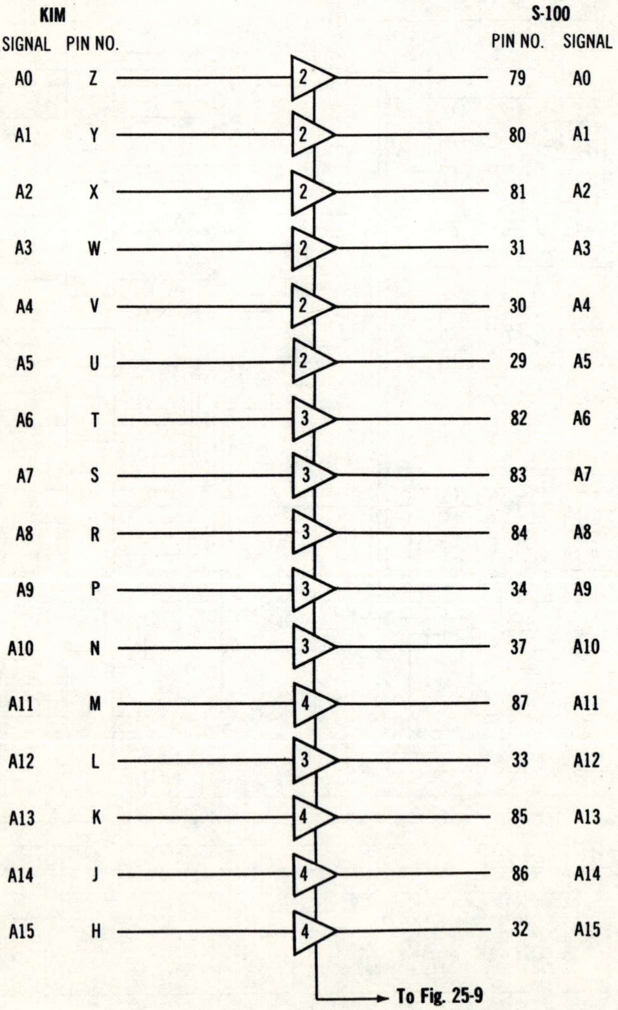

Fig. 25-3. The address lines are simply buffered lines.

Fig. 25-5 shows the derivation of the PDBIN signal, which occurs during the $\phi 2$ cycle when the 6502 is in a read operation. The R/W line of the 6502 is buffered and becomes the $\overline{\text{PWR}}$ line for the S-100. Also, the R/W line of the KIM is buffered and becomes the $\overline{\text{SWO}}$ line for the S-100. Timing of these signals is generally acceptable to most S-100 boards.

The input and output status lines are derived in Fig. 25-6. The 6502-based system treats memory and i/o devices alike: both are

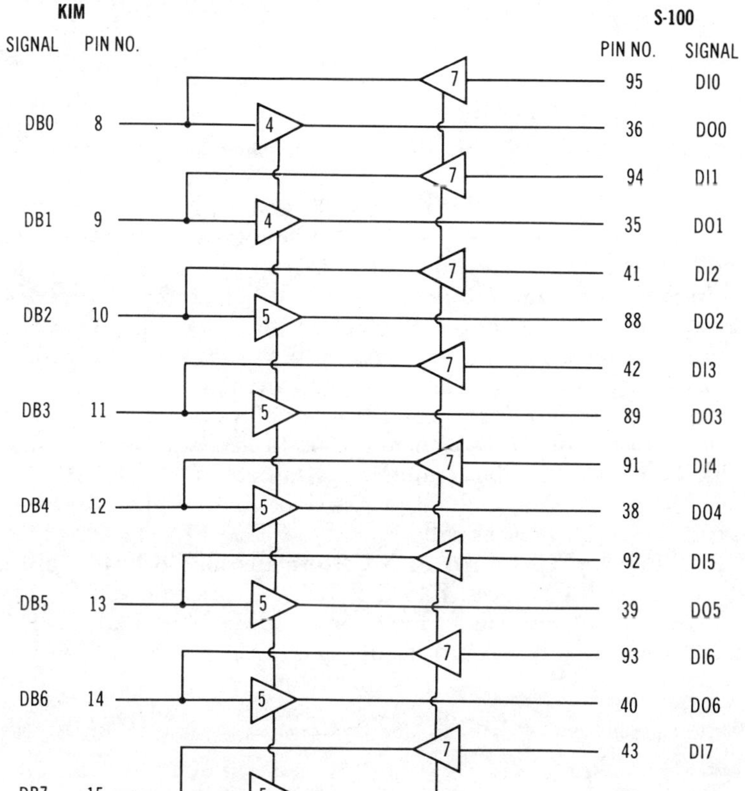

Fig. 25-4. The data lines of the KIM must be split.

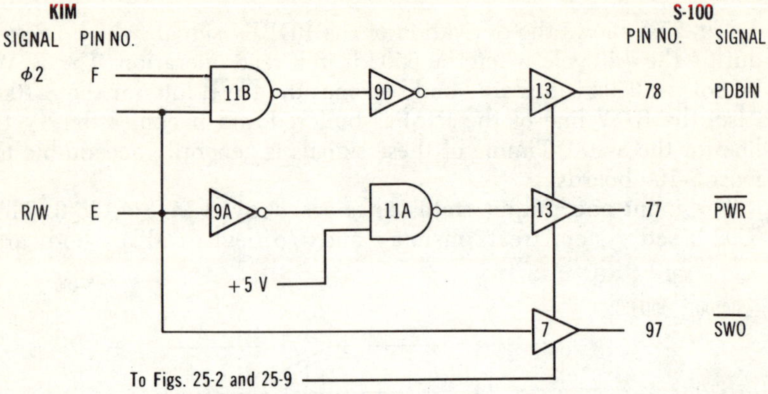

Fig. 25-5. The PDBIN, $\overline{PWR}$, and $\overline{SWO}$ lines are derived from the R/W line of the KIM.

assigned addresses between hex 0000 and FFFF and both must respond to its read/write line. The S-100 employs separate memory and i/o read/write lines, thus allowing up to 512 special i/o port addresses. The KIMSI interface board provides these special addresses by reserving one section of memory for i/o ports. These locations cannot be assigned to memory and are the only S-100 i/o locations available. When lines A12–A15 are high and line A8 is low, the decoder circuitry produces a low out of NAND gate 1A. This low is combined with the $\overline{R/W}$ of the 6502 to generate the SOUT and with the R/W of the 6502 to obtain SINP. Both SOUT and SINP will be active any time an address consisting of FYXX (where Y is even) is called. The i/o devices will respond to F0XX, F2XX, F4XX, F6XX, F8XX, FAXX, FCXX, and FEXX. Normally all i/o boards are assigned an F0XX address. F2XX, F4XX, . . . are reserved for those i/o boards that respond to the upper address lines (A8–A15). To prevent enabling two boards at the same time, a board at F000 and

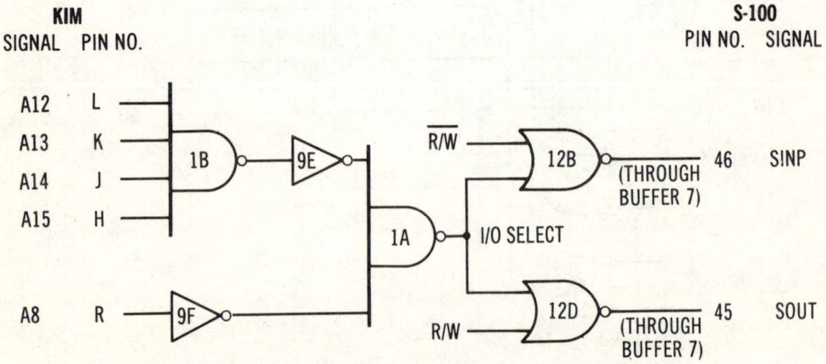

Fig. 25-6. Derivation of the input and output status lines.

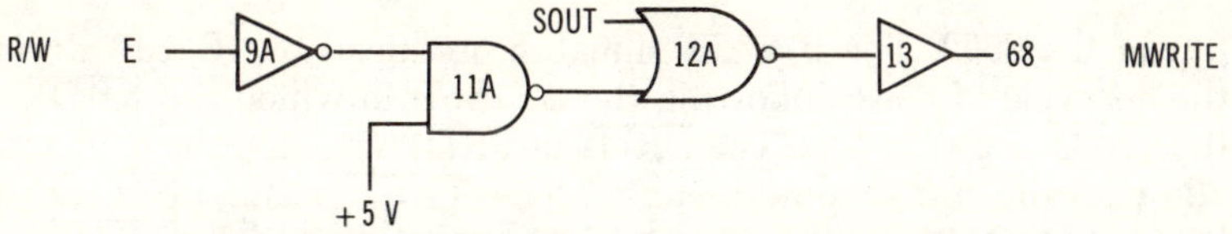

Fig. 25-7. The MWRITE is derived from the i/o select.

a board at F200, one F0XX address must be disallowed for boards using F0XX locations. Boards responding to the high byte, F2XX, etc., must use that disallowed low byte in their address. If F000, for example, is the forbidden address, conflicts will not occur as high-byte boards will be at F200, F400, etc.

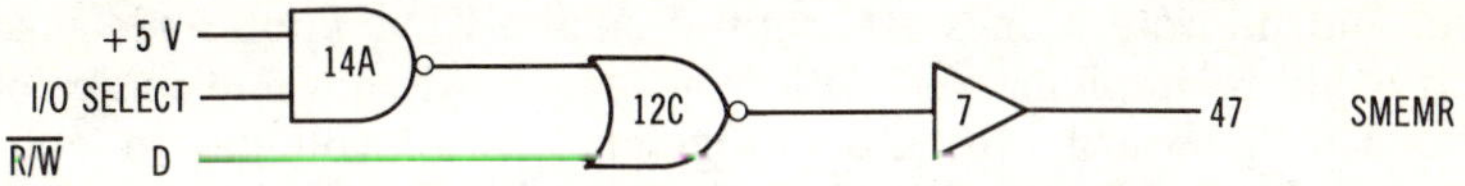

Fig. 25-8. The SMEMR is derived from the i/o select.

The MWRITE is also a product of the circuitry of Fig. 25-7. The SOUT line is NORed with the R/W of the 6502, which produces an output to memory when the i/o section has not been enabled. By combining the i/o select signal from gate 1A with the $\overline{\text{R/W}}$ of the 6502, SMEMR is obtained as shown in Fig. 25-8. By deriving the MWRITE and SMEMR from the R/W and i/o select circuitry, memory and i/o addresses are separated.

The hold and ready interface circuitry is shown in Fig. 25-9. The S-100 $\overline{\text{KRDY}}$ and $\overline{\text{PRDY}}$ lines are directly connected to the $\overline{\text{RDY}}$ in-

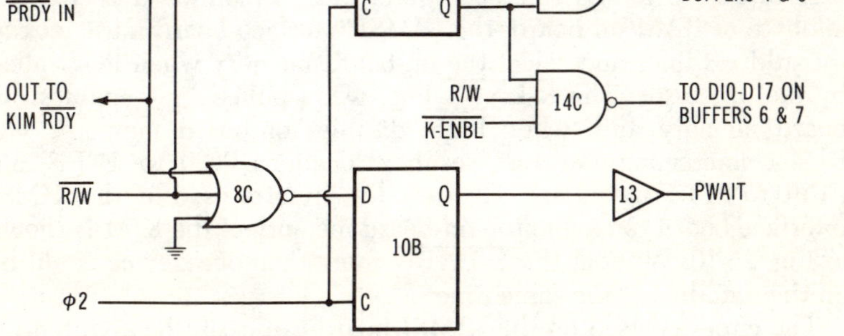

Fig. 25-9. The hold and ready interface circuitry.

put of the 6502. The PWAIT signal is produced by IC 10B during the $\phi 2$ cycle of 6502, following the $\phi 1$ cycle in which the $\overline{\text{KRDY}}$ or the $\overline{\text{PRDY}}$ is pulled low. The $\overline{\text{KRDY}}$ or $\overline{\text{PRDY}}$ will stop the 6502 from allowing the use of slow memory. The $\overline{\text{KHOLD}}$ and $\overline{\text{PHOLD}}$ can be used by DMA devices to gain control of the buses. The $\overline{\text{RDY}}$ line of the 6502 must be used in conjunction with the hold lines. Pulling the hold and ready lines low floats the address, data, and control lines during the next read cycle.

The $\overline{\text{PINT}}$ line connects directly to the $\overline{\text{IRQ}}$ input of the 6502. The $\overline{\text{NMI}}$ line of 6502 is fed directly by S-100 line 17, which is not standardized in most 8080-based systems. The $\overline{\text{POC}}$ line of the S-100 is connected to and resets the 6502 through the $\overline{\text{RST}}$ line. The protect and unprotect lines are unused on the KIMSI interface board but could be implemented with front-panel switches and indicators. The KIMSI board uses the unregulated S-100 voltages to obtain the regulated +5 V and +12 V required by the KIM-1 microcomputer.

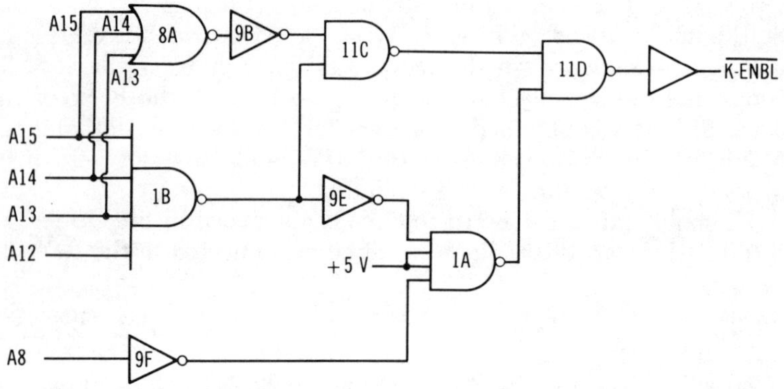

Fig. 25-10. Decoder circuitry used by the KIMSI interface board to enable the KIM-1.

Because the KIM-1 microcomputer has its monitor ROMs and 1 kilobyte of RAM on board, the KIMSI interface board must look at the address lines and select the on-board memory when it is called by the processor. The $\overline{\text{K-ENBL}}$ line, when pulled low, enables on-board memory and, when high, disables on-board memory. The KIM-1 microcomputer reserves hex locations 0000 to 1FFF and FFF0 to FFFF for its use. The decoder circuitry used by the KIMSI interface board to enable the on-board memory of the KIM is shown in Fig. 25-10. Without the $\overline{\text{K-ENBL}}$, more than one device could be on the data bus at the same time.

The circuitry used on the KIMSI board can easily be adapted for use by other 6502-based systems. Forethought Products produces a

PET to S-100 interface board, the BETSI, which is shown schematically in Fig. 25-11. Notice that although some lines are handled differently, most of the same S-100 signals are derived.

Since the 6502 and 6800 processors produce the same clock, address, data, R/W, interrupt, and reset lines, they are converted as previously discussed. The ready line is not available on the 6800 and slow memory must stretch its clock cycles. The 6800 microprocessor outputs a valid memory-address signal, VMA, each time a memory or i/o operation begins which can be used as the S-100 SYNC. The 6800 $\overline{\text{HALT}}$ input can be connected to the S-100 $\overline{\text{HOLD}}$ lines to allow DMA operations. The three-state control (TSC) of the 6800 is functionally the same as the S-100 address disable $\overline{\text{ADDR DSBL}}$. They must, however, be connected through an inverter. The S-100 data out disable ($\overline{\text{DO DSB}}$) function can be performed with the data bus enable line (DBE) of the 6800 microprocessor, if the $\overline{\text{DB DSB}}$ is inverted. The bus available (BA) output of the 6800 can be used as the S-100 hold acknowledge (PHLDA) signal, in response to a PHOLD or KHOLD input.

The conversion from the 6502/6800 buses to the S-100 bus is not difficult for most signals. Timing and signal information should be carefully examined, however, before any board is added.

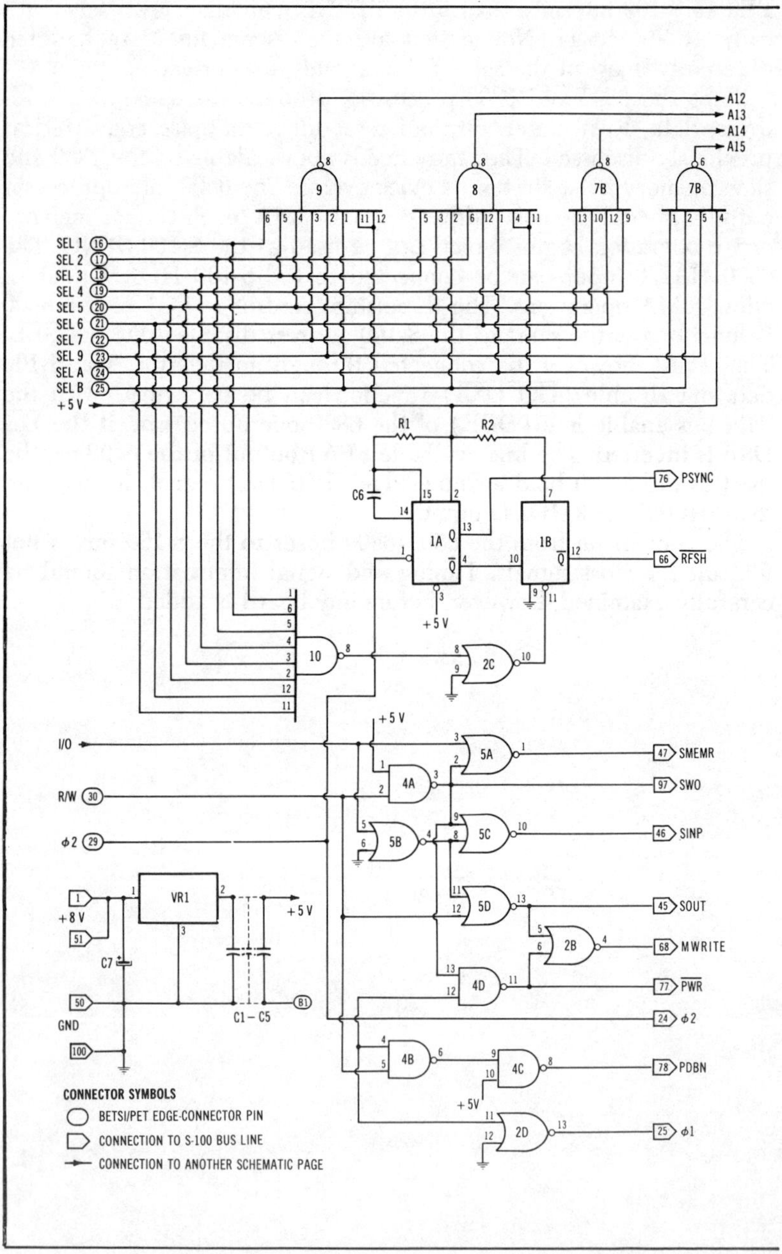

(A) Part 1.

Fig. 25-11. The schematic for the

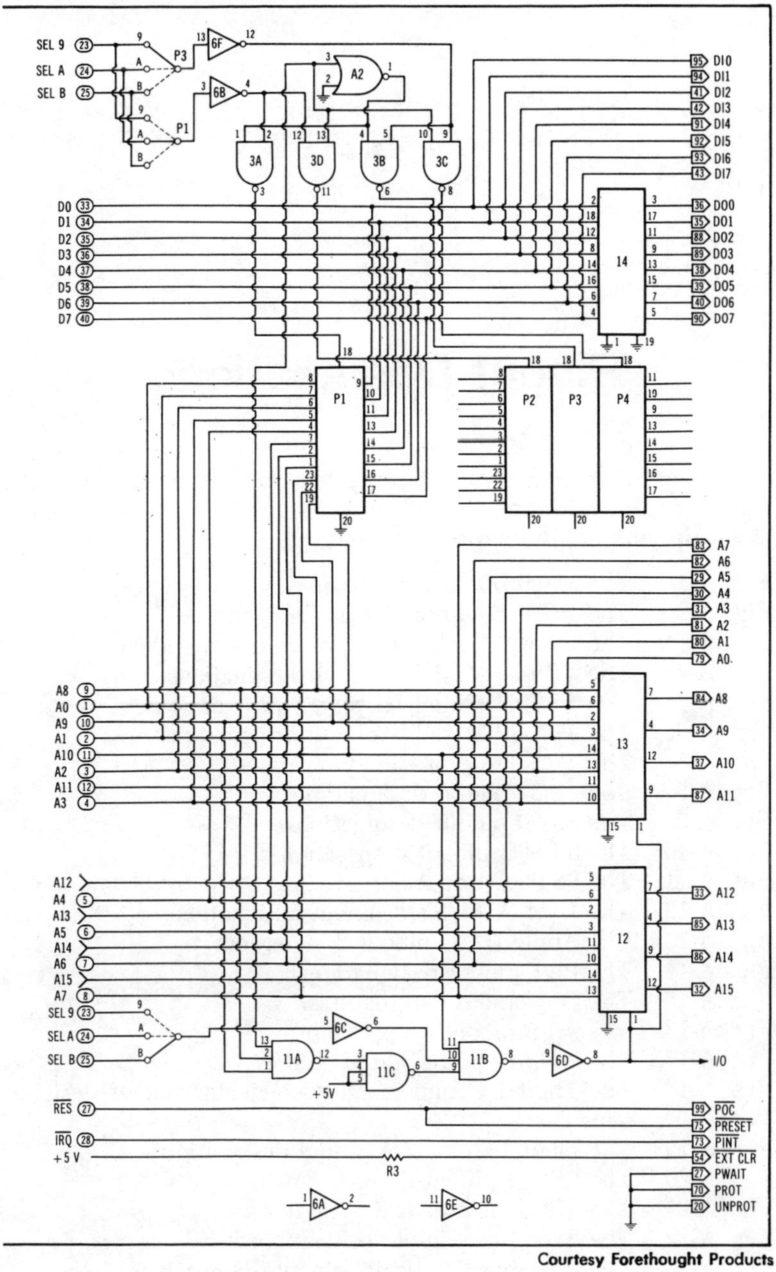

Courtesy Forethought Products

(B) Part 2.

BETSI interface board.

APPENDIX

Pinout Designations

The various pinout designations given throughout the book are listed again in this appendix. They will serve here as a convenient finger-tip guide to the pin designations.

Pin	Signal	Signal	Pin
51	+8 V	+8 V	1
52	−16 V	+16 V	2
53	$\overline{\text{SSW DSB}}$	$\overline{\text{XRDY}}$	3
54	$\overline{\text{EXT CLR}}$	VI0	4
55	*	VI1	5
56	*	VI2	6
57	*	VI3	7
58	*	VI4	8
59	*	VI5	9
60	*	VI6	10
61	*	VI7	11
62	*	*	12
63	*	*	13
64	*	*	14
65	*	*	15
66	*	*	16
67	*	*	17
68	MWRITE	$\overline{\text{STAT DSB}}$	18
69	$\overline{\text{PS}}$	$\overline{\text{C/C DSB}}$	19
70	PROT	UNPROT	20
71	RUN	SS	21
72	$\overline{\text{PRDY}}$	$\overline{\text{ADD DSB}}$	22
73	$\overline{\text{PINT}}$	$\overline{\text{DO DSB}}$	23
74	$\overline{\text{PHOLD}}$	$\phi 2$	24
75	$\overline{\text{PRESET}}$	$\phi 1$	25
76	PSYNC	PHLDA	26
77	$\overline{\text{PWR}}$	PWAIT	27
78	PDBIN	PINTE	28
79	A0	A5	29
80	A1	A4	30
81	A2	A3	31
82	A6	A15	32
83	A7	A12	33
84	A8	A9	34
85	A13	DO1	35
86	A14	DO0	36
87	A11	A10	37
88	DO2	DO4	38
89	DO3	DO5	39
90	DO7	DO6	40
91	DI4	DI2	41
92	DI5	DI3	42
93	DI6	DI7	43
94	DI1	SM1	44
95	DI0	SOUT	45
96	SINTA	SINP	46
97	$\overline{\text{SWO}}$	SMEMR	47
98	SSTACK	SHLTA	48
99	POC	$\overline{\text{CLOCK}}$	49
100	GND	GND	50

Fig. A-1. The S-100 pinout designations.

Pin	Signal	Signal	Pin
51	+8V	+8V	1
52	−16V	+16V	2
53	GND	XRDY	3
54	SLAVE CLR*	VI0*	4
55	DMA0*	VI1*	5
56	DMA1*	VI2*	6
57	DMA2*	VI3*	7
58	sXTRQ*	VI4*	8
59	A19	VI5*	9
60	SIXTN*	VI6*	10
61	A20	VI7*	11
62	A21	NMI*	12
63	A22	PWRFAIL*	13
64	A23	DMA3*	14
65	NDEF	A18	15
66	NDEF	A16	16
67	PHANTOM*	A17	17
68	MWRT	SDSB*	18
69	RFU	CDSB*	19
70	GND	GND	20
71	NDEF	RFU	21
72	RDY	ADSB*	22
73	INT*	DODSB*	23
74	HOLD*	ϕ	24
75	RESET	pSTVAL*	25
76	pSYNC	pHLDA	26
77	pWR*	RFU	27
78	pDBIN	RFU	28
79	A0	A5	29
80	A1	A4	30
81	A2	A3	31
82	A6	A15	32
83	A7	A12	33
84	A8	A9	34
85	A13	DO1/DATA1	35
86	A14	DO0/DATA0	36
87	A11	A10	37
88	DO2/DATA2	DO4/DATA4	38
89	DO3/DATA3	DO5/DATA5	39
90	DO7/DATA7	DO6/DATA6	40
91	DI4/DATA12	DI2/DATA10	41
92	DI5/DATA13	DI3/DATA11	42
93	DI6/DATA14	DI7/DATA15	43
94	DI1/DATA9	sM1	44
95	DI0/DATA8	sOUT	45
96	sINTA	sINP	46
97	sWO*	sMEMR	47
98	ERROR*	sHLTA	48
99	POC*	CLOCK	49
100	GND	GND	50

Fig. A-2. The IEEE S-100 pinout designations.

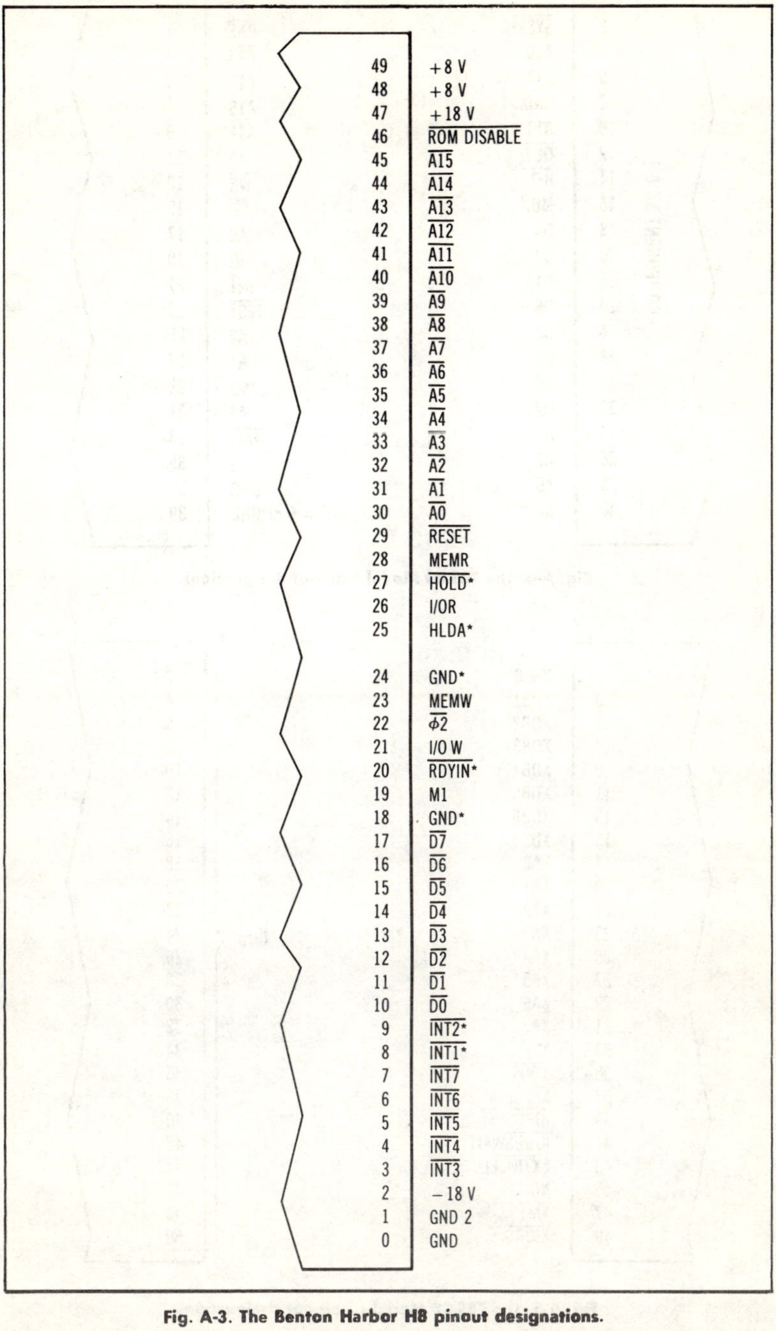

Fig. A-3. The Benton Harbor H8 pinout designations.

COMPONENT SIDE

Pin	Signal	Signal	Pin
2	$\overline{\text{SYSRES}}$	$\overline{\text{RAS}}$	1
4	A10	$\overline{\text{CAS}}$	3
6	A13	A12	5
8	GND	A15	7
10	A14	A11	9
12	$\overline{\text{OUT}}$	A8	11
14	$\overline{\text{INTAK}}$	$\overline{\text{WR}}$	13
16	MUX	$\overline{\text{RD}}$	15
18	D4	A9	17
20	D7	$\overline{\text{IN}}$	19
22	D1	$\overline{\text{INT}}$	21
24	D6	$\overline{\text{TEST}}$	23
26	D3	A0	25
28	D5	A1	27
30	D0	GND	29
32	D2	A4	31
34	A3	$\overline{\text{WAIT}}$	33
36	A7	A5	35
38	A6	GND	37
40	A2	+5 or GND	39

Fig. A-4. The TRS-80 Model I pinout designations.

Pin	Signal	Signal	Pin
1	XBC0	GND	2
3	XDB1	GND	4
5	XDB2	GND	6
7	XDB3	GND	8
9	XDB4	GND	10
11	XDB5	GND	12
13	XDB6	GND	14
15	XDB7	GND	16
17	XA0	GND	18
19	XA1	GND	20
21	XA2	GND	22
23	XA3	GND	24
25	XA4	GND	26
27	XA5	GND	28
29	XA6	GND	30
31	XA7	GND	32
33	$\overline{\text{XIN}}$	GND	34
35	$\overline{\text{XOUT}}$	GND	36
37	$\overline{\text{XRESET}}$	GND	38
39	$\overline{\text{IOBUSINT}}$	GND	40
41	$\overline{\text{IOBUSWAIT}}$	GND	42
43	$\overline{\text{EXTIOSEL}}$	GND	44
45	NC	GND	46
47	$\overline{\text{XM1}}$	GND	48
49	$\overline{\text{XIORQ}}$	GND	50

Fig. A-5. The TRS-80 Model III pinout designations.

Pin	Signal	Signal	Pin
1	USER0	USER1	2
3	GND	GND	4
5	+12V	+12V	6
7	GND	GND	8
9	+5V	+5V	10
11	$\overline{INTRQ}$	$\overline{NMIRQ}$	12
13	IEIN	IEOUT	14
15	$\overline{BAKIN}$	$\overline{BAKOUT}$	16
17	$\overline{BUSRQ}$	$\overline{SYNC}$	18
19	$\overline{RD}$	$\overline{WR}$	20
21	$\overline{MEMCYC}$	$\overline{IOCYC}$	22
23	$\overline{A00}$	$\overline{A01}$	24
25	$\overline{A02}$	$\overline{A03}$	26
27	$\overline{A04}$	$\overline{A05}$	28
29	$\overline{A06}$	$\overline{A07}$	30
31	$\overline{A08}$	$\overline{A09}$	32
33	$\overline{A10}$	$\overline{A11}$	34
35	$\overline{A12}$	$\overline{A13}$	36
37	$\overline{A14}$	$\overline{A15}$	38
39	RES	$\overline{DISRO}$	40
41	XFERRQ	$\overline{KBIRQ}$	42
43	$\overline{SELECT}$	CLOCK	44
45	$\overline{REFRSH}$	8MHz	46
47	RTC	$\overline{WAIT}$	48
49	GND	GND	50
51	$\overline{DAT0}$	$\overline{DAT1}$	52
53	$\overline{DAT2}$	$\overline{DAT3}$	54
55	$\overline{DAT4}$	$\overline{DAT5}$	56
57	$\overline{DAT6}$	$\overline{DAT7}$	58
59	RES	RES	60
61	RES	RES	62
63	RES	RES	64
65	RES	RES	66
67	$\overline{RESET}$	$\overline{HALT}$	68
69	GND	GND	70
71	+5V	+5V	72
73	GND	GND	74
75	−12V	−12V	76
77	+12V	+12V	78
79	GND	GND	80

Fig. A-6. The TRS-80 Model II pinout designations.

Pin	Signal	Signal	Pin
1	−12V	+12V	2
3	$\overline{HALT}$	$\overline{NMI}$	4
5	$\overline{RESET}$	E	6
7	Q	$\overline{CART}$	8
9	+5V	D0	10
11	D1	D2	12
13	D3	D4	14
15	D5	D6	16
17	D7	$R/\overline{W}$	18
19	A0	A1	20
21	A2	A3	22
23	A4	A5	24
25	A6	A7	26
27	A8	A9	28
29	A10	A11	30
31	A12	$\overline{CTS}$	32
33	GND	GND	34
35	SND	$\overline{SCS}$	36
37	A13	A14	38
39	A15	SLEND	40

Fig. A-7. The TRS-80 Color Computer pinout designations.

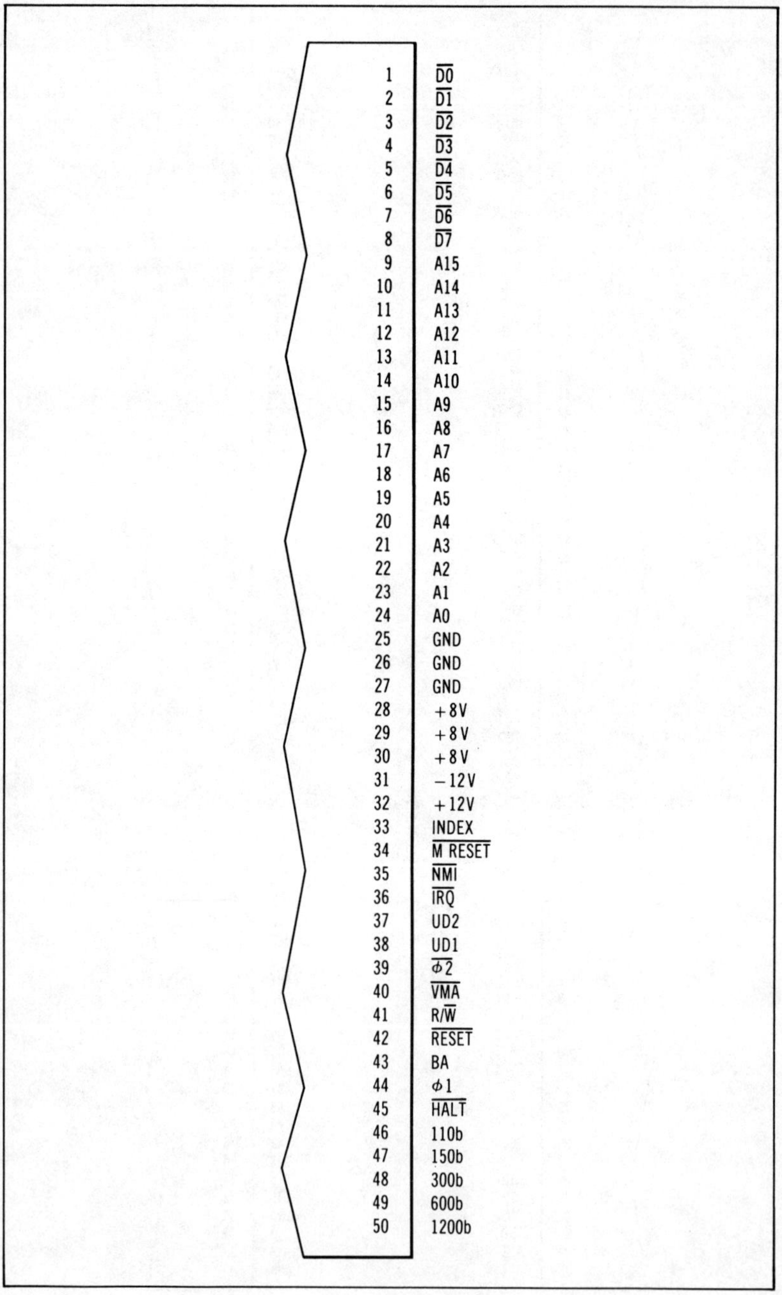

Fig. A-8. The SS-50 pinout designations.

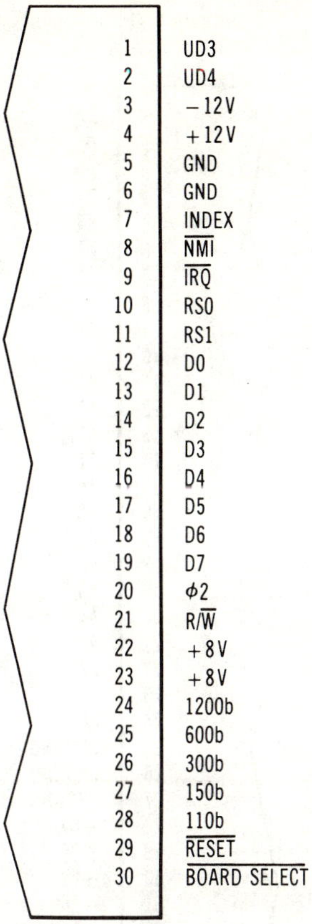

Fig. A-9. The SS-30 pinout designations.

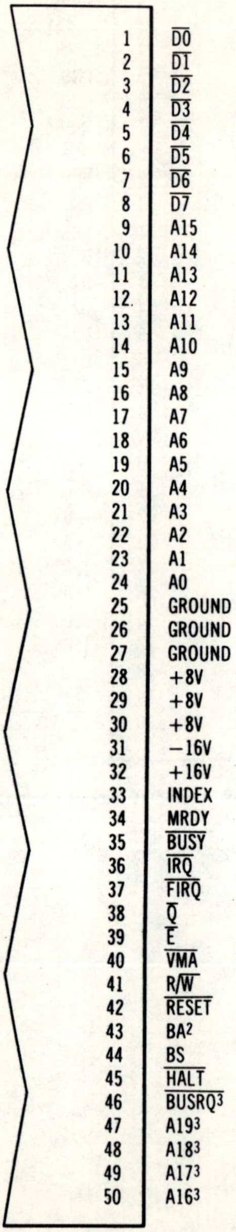

Fig. A-10. The SS-50C pinout designations.

Pin	Signal
1	RS2
2	RS3
3	−16V
4	+16V
5	GND
6	GND
7	$\overline{\text{INDEX}}$
8	$\overline{\text{FIRQ}}$
9	$\overline{\text{IRQ}}$
10	RS0
11	RS1
12	D0
13	D1
14	D2
15	D3
16	D4
17	D5
18	D6
19	D7
20	$\overline{\text{E}}$
21	R/$\overline{\text{W}}$
22	+8V
23	+8V
24	1200b
25	4800b
26	300b
27	9600b
28	110b
29	$\overline{\text{RESET}}$
30	$\overline{\text{I/O SEL}}$

Fig. A-11. The SS-30C pinout designations.

Pin	Signal	Signal	Pin
A	GND	GND	1
B	BAB0	BSYNC	2
C	BAB1	$\overline{\text{BRDY}}$	3
D	BAB2	$\overline{\text{BIRQ}}$	4
E	BAB3	−15 V	5
F	BAB4	$\overline{\text{BNMI}}$	6
H	BAB5	$\overline{\text{BRST}}$	7
J	BAB6	BDB7	8
K	BAB7	BDB6	9
L	BAB8	BDB5	10
M	BAB9	BDB4	11
N	BAB10	BDB3	12
P	BAB11	BDB2	13
R	BAB12	DBD1	14
S	BAB13	BDB0	15
T	BAB14	BD SEL	16
U	BAB15	+15 V	17
V	BΦ2	$\overline{\text{DMA}}$	18
W	BR/W	+8 V	19
X	B$\overline{\Phi 2}$	+8 V	20
Y	+5 V	+5 V	21
Z	GND	GND	22

Fig. A-12. The KIM/AIM/SYM pinout designations.

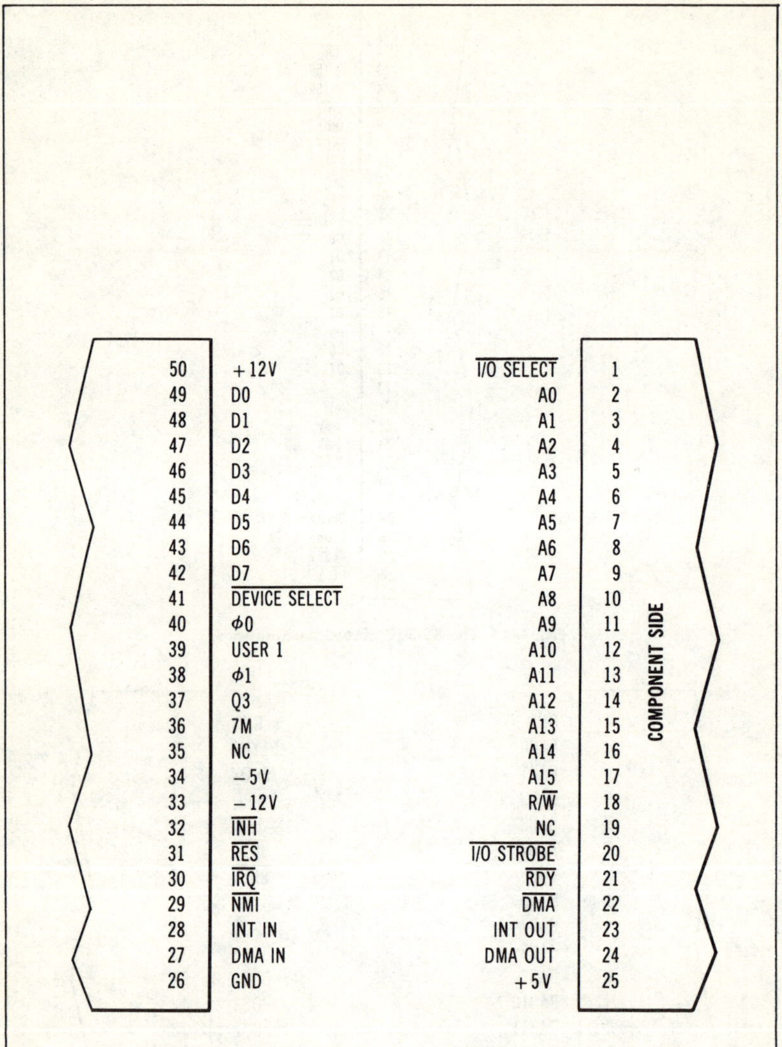

Fig. A-13. The Apple II/III pinout designations.

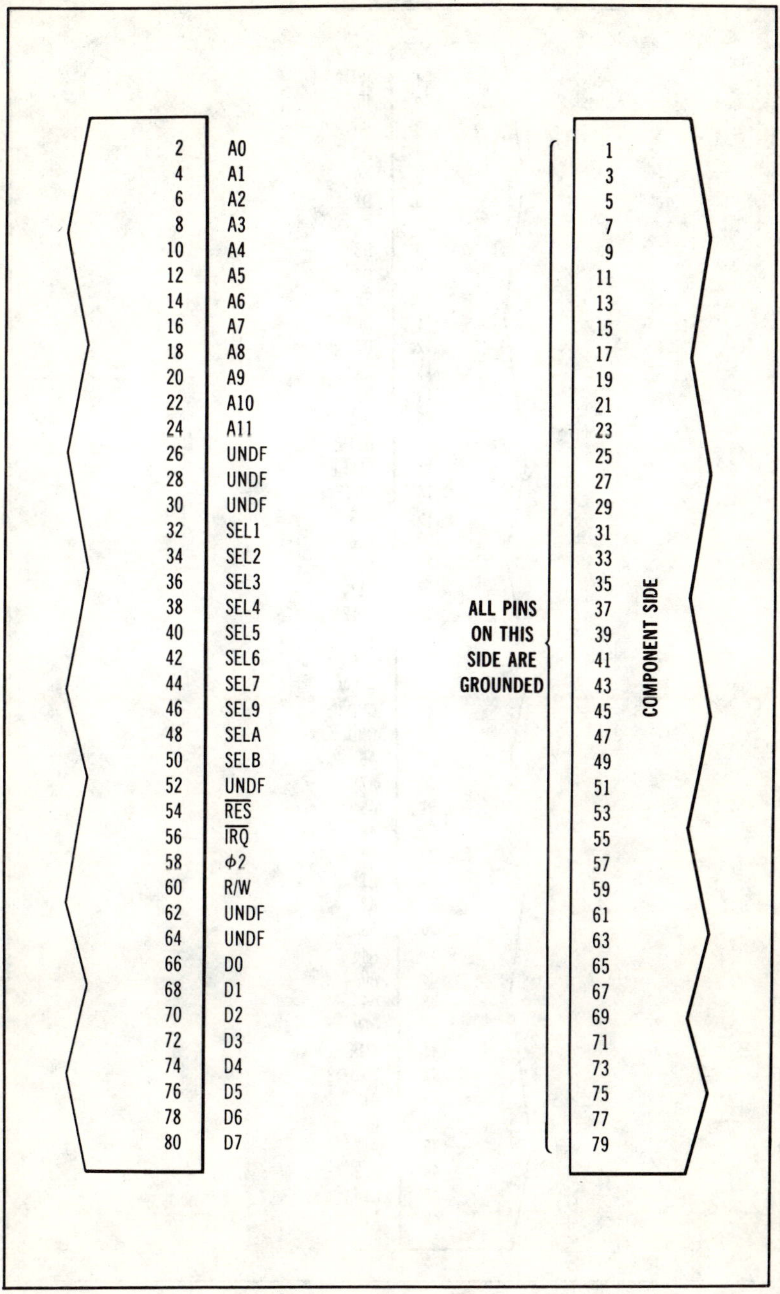

Fig. A-14. The PET pinout designations.

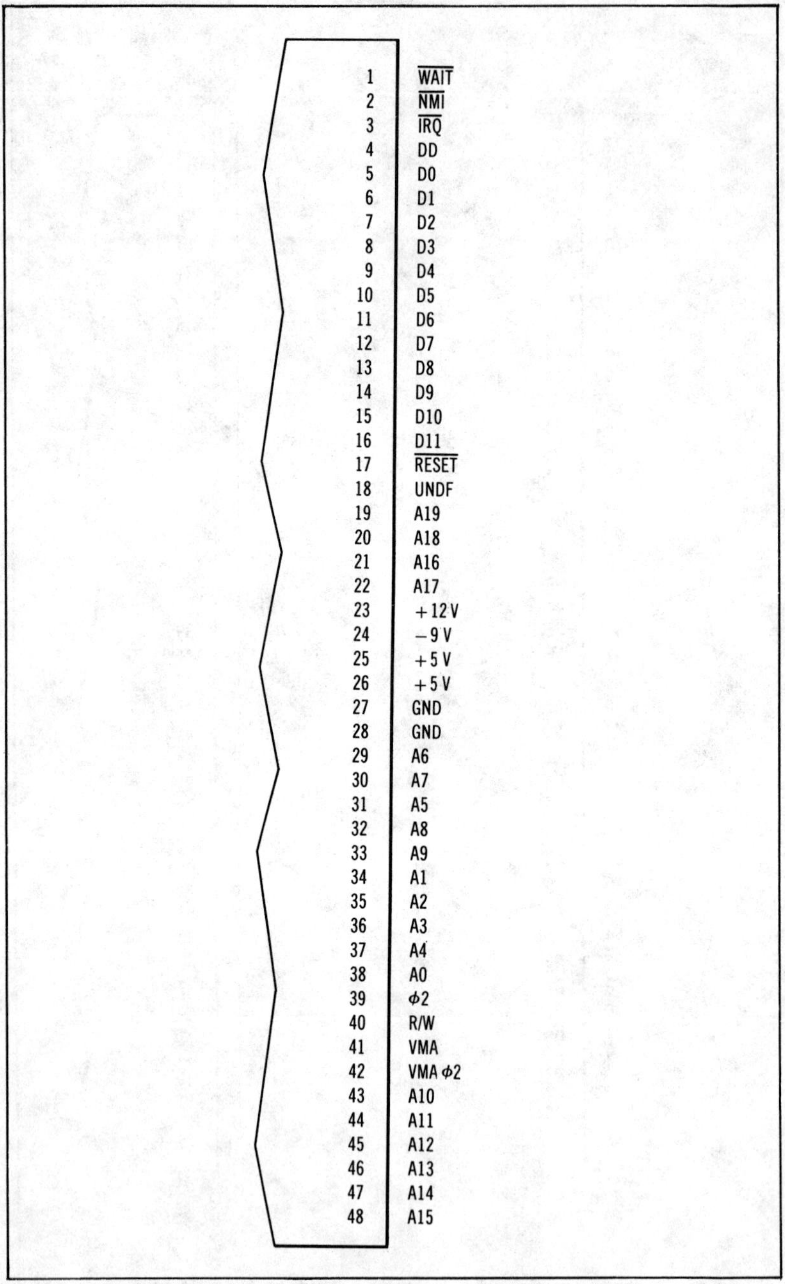

Fig. A-15. The OSI pinout designations.

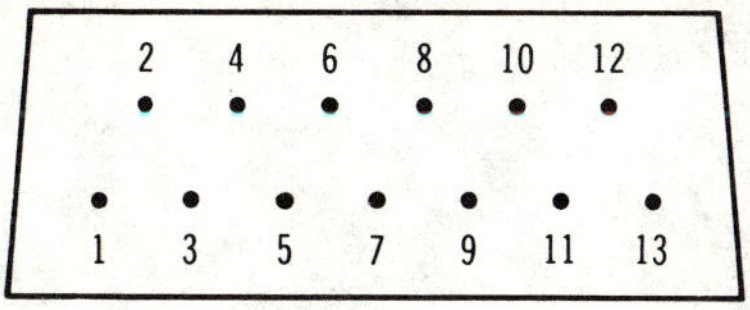

1 CLOCK INPUT
2 CLOCK OUTPUT
3 DATA INPUT
4 GROUND
5 DATA OUTPUT
6 GROUND
7 COMMAND
8 MOTOR CONTROL
9 PROCEED
10 +5/READY
11 AUDIO INPUT
12 +12V
13 INTERRUPT

Fig. A-16. The Atari pinout designations.

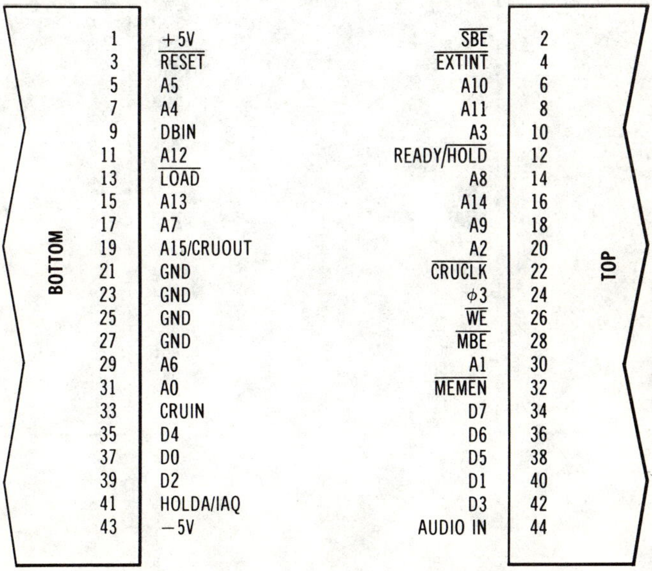

Fig. A-17. The TI-99/4 pinout designations.

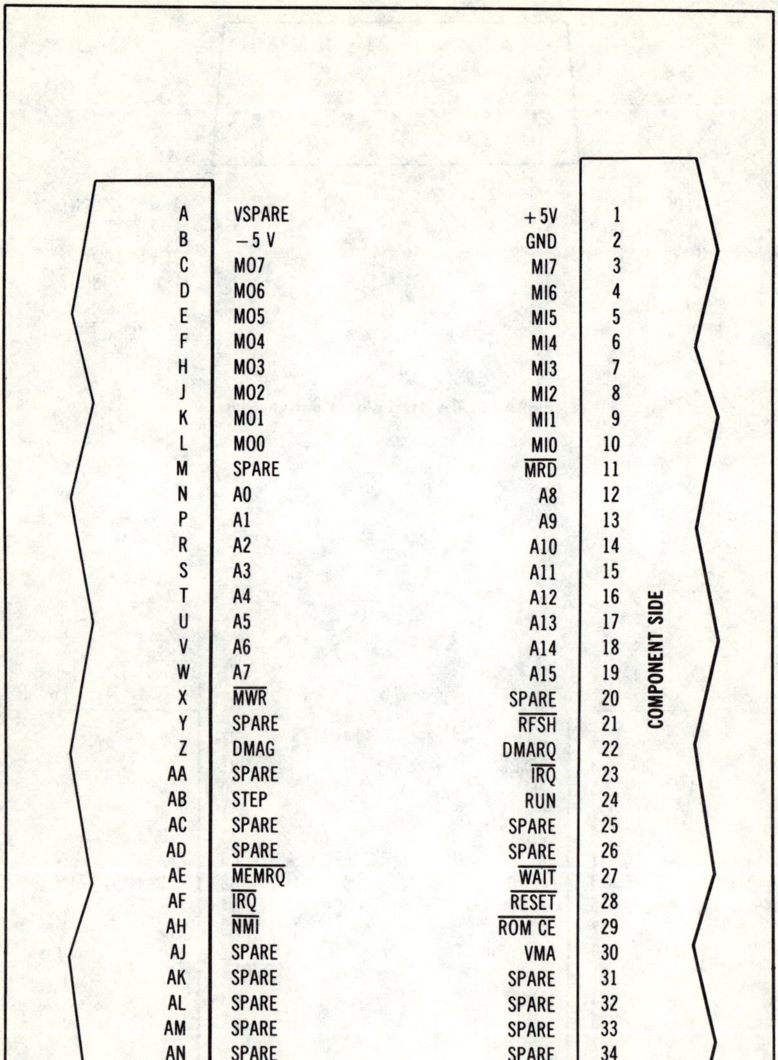

Fig. A-18. The Digital Group memory-connector pinout designations.

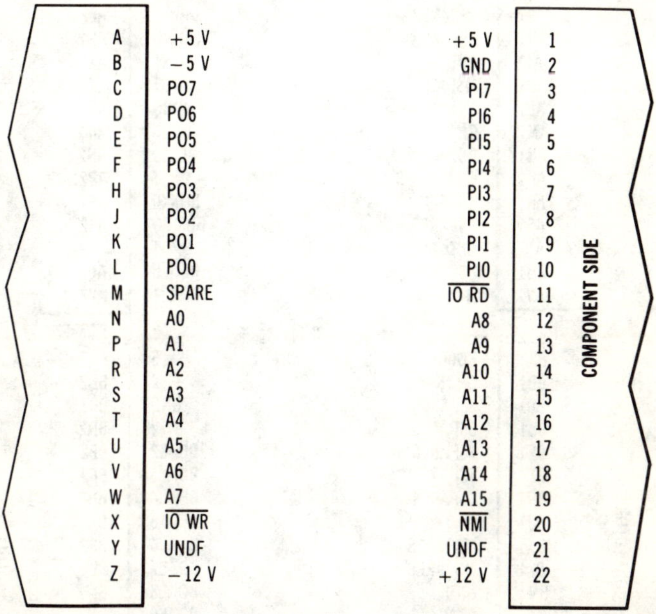

Fig. A-19. The Digital Group i/o pinout designations.

COMPONENT SIDE / SOLDER SIDE

CONNECTOR A

Pin (Component Side)	Signal	Signal	Pin (Solder Side)
AA1	BSPARE1	+5	AA2
AB1	BSPARE2	−12	AB2
AC1	BAD16	GND	AC2
AD1	BAD17	+12	AD2
AE1	SSPARE1	BDOUT	AE2
AF1	SSPARE2	BRPLY	AF2
AH1	SSPARE3	BDIN	AH2
AJ1	GND	BSYNC	AJ2
AK1	MSPAREA	BWTBT	AK2
AL1	MSPAREA	BIRQ	AL2
AM1	GND	BIAKI	AM2
AN1	BDMR	BIAKO	AN2
AP1	BHALT	BBS7	AP2
AR1	BREF	BDMGI	AR2
AS1	PSPARE3	BDMGO	AS2
AT1	GND	BINIT	AT2
AU1	PSPARE1	BDAL0	AU2
AV1	+5B	BDAL1	AV2

CONNECTOR B

Pin (Component Side)	Signal	Signal	Pin (Solder Side)
BA1	BDCOK	+5	BA2
BB1	DPOK	−12	BB2
BC1	SSPARE4	GND	BC2
BD1	SSPARE5	+12	BD2
BE1	SSPARE6	BDAL2	BE2
BF1	SSPARE7	BDAL3	BF2
BH1	SSPARE8	BDAL4	BH2
BJ1	GND	BDAL5	BJ2
BK1	MSPAREB	BDAL6	BK2
BL1	MSPAREB	BDAL7	BL2
BM1	GND	BDAL8	BM2
BN1	BSACK	BDAL9	BN2
BP1	BSPARE6	BDAL10	BP2
BR1	BEVNT	BDAL11	BR2
BS1	PSPARE4	BDAL12	BS2
BT1	GND	BDAL13	BT2
BU1	PSPARE2	BDAL14	BU2
BV1	+5	BDAL15	BV2

Fig. A-20. The LSI-11 pinout designations.

Pin	Signal	Signal	Pin
2	GND	GND	1
4	+5 V	+5 V	3
6	+5 V	+5 V	5
8	+12 V	+12 V	7
10	−5 V	−5 V	9
12	GND	GND	11
14	$\overline{\text{INIT}}$	$\overline{\text{BCLK}}$	13
16	$\overline{\text{BPRO}}$	$\overline{\text{BPRN}}$	15
18	$\overline{\text{BREQ}}$	$\overline{\text{BUSY}}$	17
20	$\overline{\text{MWTC}}$	$\overline{\text{MRDC}}$	19
22	$\overline{\text{IOWC}}$	$\overline{\text{IORC}}$	21
24	$\overline{\text{INH1}}$	$\overline{\text{XACK}}$	23
26	$\overline{\text{INH2}}$	$\overline{\text{AACK}}$	25
28	UNDF	UNDF	27
30	UNDF	UNDF	29
32	UNDF	$\overline{\text{CCLK}}$	31
34	UNDF	$\overline{\text{INTR}}$	33
36	$\overline{\text{INT7}}$	$\overline{\text{INT6}}$	35
38	$\overline{\text{INT5}}$	$\overline{\text{INT4}}$	37
40	$\overline{\text{INT3}}$	$\overline{\text{INT2}}$	39
42	$\overline{\text{INT1}}$	$\overline{\text{INT0}}$	41
44	$\overline{\text{ADRF}}$	$\overline{\text{ADRE}}$	43
46	$\overline{\text{ADRD}}$	$\overline{\text{ADRC}}$	45
48	$\overline{\text{ADRB}}$	$\overline{\text{ADRA}}$	47
50	$\overline{\text{ADR9}}$	$\overline{\text{ADR8}}$	49
52	$\overline{\text{ADR7}}$	$\overline{\text{ADR6}}$	51
54	$\overline{\text{ADR5}}$	$\overline{\text{ADR4}}$	53
56	$\overline{\text{ADR3}}$	$\overline{\text{ADR2}}$	55
58	$\overline{\text{ADR1}}$	$\overline{\text{ADR0}}$	57
60	$\overline{\text{DATF}}$	$\overline{\text{DATE}}$	59
62	$\overline{\text{DATD}}$	$\overline{\text{DATC}}$	61
64	$\overline{\text{DATB}}$	$\overline{\text{DATA}}$	63
66	$\overline{\text{DAT9}}$	$\overline{\text{DAT8}}$	65
68	$\overline{\text{DAT7}}$	$\overline{\text{DAT6}}$	67
70	$\overline{\text{DAT5}}$	$\overline{\text{DAT4}}$	69
72	$\overline{\text{DAT3}}$	$\overline{\text{DAT2}}$	71
74	$\overline{\text{DAT1}}$	$\overline{\text{DAT0}}$	73
76	GND	GND	75
78	−10 V	−10 V	77
80	−12 V	−12 V	79
82	+5 V	+5 V	81
84	+5 V	+5 V	83
86	GND	GND	85

COMPONENT SIDE

Fig. A-21. The Multibus pinout designations.

Pin	Signal	Signal	Pin
1	+5V	+5V	A
2	+5V	+5V	B
3	+5V	+5V	C
4	GO/HLT	$\overline{\text{IRQ}}$	D
5	$\overline{\text{RESET}}$	$\overline{\text{NMI}}$	E
6	R/W	VMA	F
7	ϕ1	GND	H
8	GND	ϕ2	J
9	GND	GND	K
10	VUA	MEM CLK	L
11	−12V	−12V	M
12	$\overline{\text{REF REQ}}$	TSC	N
13	REF GNT	BA	P
14	UNDF	MEM RDY	R
15	UNDF	REF CLK	S
16	+12V	+12V	T
17	UNDF	UNDF	U
18	UNDF	UNDF	V
19	UNDF	UNDF	W
20	UNDF	UNDF	X
21	UNDF	UNDF	Y
22	UNDF	UNDF	Z
23	UNDF	UNDF	$\overline{\text{A}}$
24	UNDF	UNDF	$\overline{\text{B}}$
25	UNDF	UNDF	$\overline{\text{C}}$
26	UNDF	UNDF	$\overline{\text{D}}$
27	UNDF	UNDF	$\overline{\text{E}}$
28	UNDF	UNDF	$\overline{\text{F}}$
29	$\overline{\text{D1}}$	$\overline{\text{D3}}$	$\overline{\text{H}}$
30	$\overline{\text{D5}}$	$\overline{\text{D7}}$	$\overline{\text{J}}$
31	$\overline{\text{D0}}$	$\overline{\text{D2}}$	$\overline{\text{K}}$
32	$\overline{\text{D4}}$	$\overline{\text{D6}}$	$\overline{\text{L}}$
33	A15	A14	$\overline{\text{M}}$
34	A12	A13	$\overline{\text{N}}$
35	A11	A10	$\overline{\text{P}}$
36	A8	A9	$\overline{\text{R}}$
37	A7	A6	$\overline{\text{S}}$
38	A4	A5	$\overline{\text{T}}$
39	A3	A2	$\overline{\text{U}}$
40	A0	A1	$\overline{\text{V}}$
41	GND	GND	$\overline{\text{W}}$
42	GND	GND	$\overline{\text{X}}$
43	GND	GND	$\overline{\text{Y}}$

Fig. A-22. The Exorcisor pinout designations.

Component Side Pin	Component Side Signal	Solder Side Signal	Solder Side Pin
1	+5V	+5V	2
3	GND	GND	4
5	D00	D01	6
7	D02	D03	8
9	D04	D05	10
11	D06	D07	12
13	D08	D09	14
15	D10	D11	16
17	D12	D13	18
19	D14	D15	20
21	DPARITY0	DPARITY1	22
23	GND	GND	24
25	DS0	DS1	26
27	GND	GND	28
29	DTACK	AS	30
31	GND	GND	32
33	APARITY0	WRITE	34
35	RESERVED	A01	36
37	A02	A03	38
39	A04	A05	40
41	A06	A07	42
43	A08	A09	44
45	A10	A11	46
47	A12	A13	48
49	A14	A15	50
51	A16	A17	52
53	A18	A19	54
55	A20	A21	56
57	A22	A23	58
59	IOEN	EXTDADR	60
61	GND	GND	62
63	SECMAP	ROMDIS	64
65	PDOWN	RAMDIS	66
67	GND	GND	68
69	ACCLK	SYSCLK	70
71	GND	GND	72
73	SECRESET	SYSRESET	74
75	SYSPAUSE	SYSPACK	76
77	SECIEN	ACFAIL	78
79	TEST	SYSFAIL	80
81	BERR	RETRY	82
83	FC0	FC1	84
85	FC2	FC3	86
87	IRQ1	IRQ2	88
89	IRQ3	IRQ4	90
91	IRQ5	IRQ6	92
93	IRQ7	IACK	94
95	IACKIN	IACKOUT	96
97	BG0IN	BG0OUT	98
99	BG1IN	BG1OUT	100
101	BG2IN	BG2OUT	102
103	BG3IN	BG3OUT	104
105	BG4IN	BG4OUT	106
107	BR0	BR1	108
109	BR2	BR3	110
111	BR4	BBSY	112
113	BCLR	BREL	114
115	RESERVED	RESERVED	116
117	RESERVED	RESERVED	118
119	GND	GND	120
121	−12V	−12V	122
123	GND	GND	124
125	+12V	+12V	126
127	+12V	+12V	128
129	+5V	+5V	130
131	+5V	+5V	132
133	+5STDBY	+5STDBY	134
135	GND	GND	136
137	GND	GND	138
139	GND	GND	140

Fig. A-23. The Versabus P1 and P2 pinout designations.

Component Side			Solder Side
1	GND	GND	2
3	GND	GND	4
5	GND	GND	6
7	+5V	+5V	8
9	+5V	+5V	10
11	+12V	+12V	12
13	GND	GND	14
15	−12V	−12V	16
17	I/O NOT BUSED	I/O NOT BUSED	18
19	I/O NOT BUSED	I/O NOT BUSED	20
21	I/O NOT BUSED	I/O NOT BUSED	22
23	I/O NOT BUSED	I/O NOT BUSED	24
25	I/O NOT BUSED	I/O NOT BUSED	26
27	I/O NOT BUSED	I/O NOT BUSED	28
29	I/O NOT BUSED	I/O NOT BUSED	30
31	I/O NOT BUSED	I/O NOT BUSED	32
33	I/O NOT BUSED	I/O NOT BUSED	34
35	I/O NOT BUSED	I/O NOT BUSED	36
37	I/O NOT BUSED	I/O NOT BUSED	38
39	I/O NOT BUSED	I/O NOT BUSED	40
41	I/O NOT BUSED	I/O NOT BUSED	42
43	I/O NOT BUSED	I/O NOT BUSED	44
45	I/O NOT BUSED	I/O NOT BUSED	46
47	I/O NOT BUSED	I/O NOT BUSED	48
49	I/O NOT BUSED	I/O NOT BUSED	50
51	I/O NOT BUSED	I/O NOT BUSED	52
53	I/O NOT BUSED	I/O NOT BUSED	54
55	I/O NOT BUSED	I/O NOT BUSED	56
57	I/O NOT BUSED	I/O NOT BUSED	58
59	I/O NOT BUSED	I/O NOT BUSED	60
61	I/O NOT BUSED	I/O NOT BUSED	62
63	I/O NOT BUSED	I/O NOT BUSED	64
65	I/O NOT BUSED	I/O NOT BUSED	66
67	−15V	−15V	68
69	+15V	+15V	70
71	SERCOM 1	SERCOM 2	72
73	SERCOM 3	SERCOM 4	74
75	RESERVED	RESERVED	76
77	RESERVED	RESERVED	78
79	RESERVED	RESERVED	80
81	RESERVED	RESERVED	82
83	RESERVED	RESERVED	84
85	RESERVED	RESERVED	86
87	RESERVED	$\overline{\text{APARITY1}}$	88
89	$\overline{\text{A23}}$	$\overline{\text{A25}}$	90
91	$\overline{\text{A26}}$	$\overline{\text{A27}}$	92
93	$\overline{\text{A28}}$	$\overline{\text{A29}}$	94
95	$\overline{\text{A30}}$	$\overline{\text{A31}}$	96
97	GND	GND	98
99	$\overline{\text{DS2}}$	$\overline{\text{DS3}}$	100
101	GND	GND	102
103	$\overline{\text{DPARITY2}}$	$\overline{\text{DPARITY3}}$	104
105	$\overline{\text{D16}}$	$\overline{\text{D17}}$	106
107	$\overline{\text{D18}}$	$\overline{\text{D19}}$	108
109	$\overline{\text{D20}}$	$\overline{\text{D21}}$	110
111	$\overline{\text{D22}}$	$\overline{\text{D23}}$	112
113	$\overline{\text{D24}}$	$\overline{\text{D25}}$	114
115	$\overline{\text{D26}}$	$\overline{\text{D27}}$	116
117	$\overline{\text{D28}}$	$\overline{\text{D29}}$	118
119	$\overline{\text{D30}}$	$\overline{\text{D31}}$	120

Fig. A-23. The Versabus P1 and P2 pinout designations—(continued).

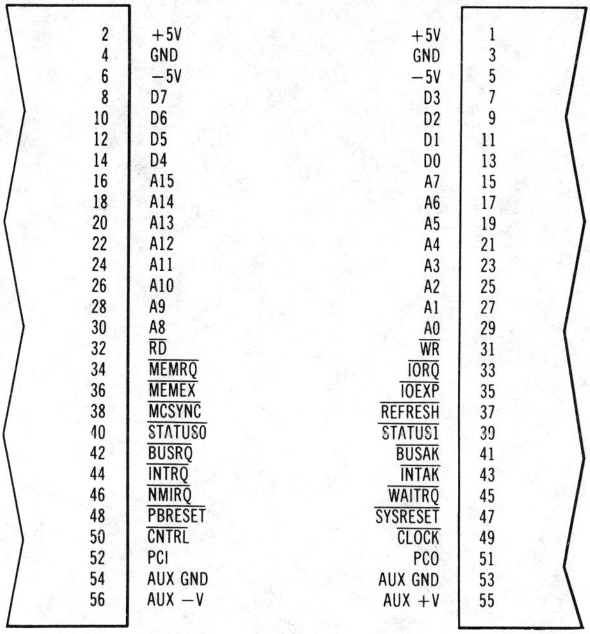

Fig. A-24. The STD pinout designations.

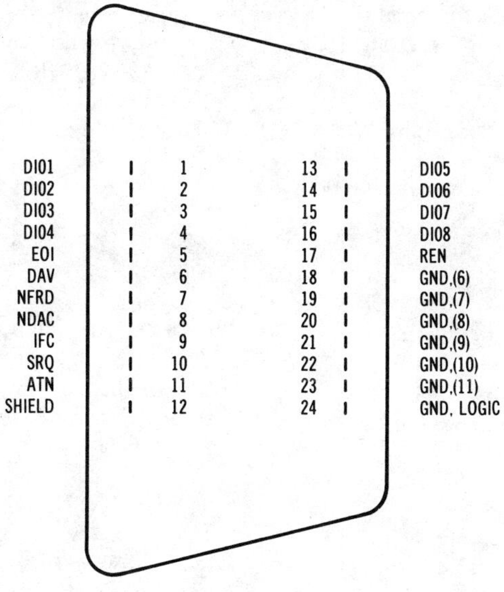

Fig. A-25. The IEEE 488 pinout designations.

Index

R

S

T

V

TO THE READER

Sams Computer books cover Fundamentals — Programming — Interfacing — Technology written to meet the needs of computer engineers, professionals, scientists, technicians, students, educators, business owners, personal computerists and home hobbyists.

Our Tradition is to meet your needs and in so doing we invite you to tell us what your needs and interests are by completing the following:

1. I need books on the following topics:

____

____

____

____

____

____

2. I have the following Sams titles:

____

____

____

____

____

____

3. My occupation is:

_____ Scientist, Engineer

_____ Personal computerist

_____ Technician, Serviceman

_____ Educator

_____ Student

_____ D P Professional

_____ Business owner

_____ Computer store owner

_____ Home hobbyist

Other ______________________

____

Name (print) ______________________________

Address ______________________________

City ____________________ State ____________ Zip ____________

Mail to: **Howard W. Sams & Co., Inc.**
Marketing Dept. #CBS1/80
4300 W. 62nd St., P.O. Box 7092
Indianapolis, Indiana 46206

21810